Practical *in situ* Hybridization

Practical *in situ* Hybridization

Trude Schwarzacher and Pat Heslop-Harrison
*Departments of Cell Biology and Cereals Research, John Innes Centre,
Norwich, UK*

First published 2000

A CIP catalogue record for this book is available from the British Library.

ISBN 1 85996 138 X

BIOS Scientific Publishers Ltd
9 Newtec Place, Magdalen Road, Oxford OX4 1RE, UK
Tel. +44 (0)1865 726286. Fax +44 (0)1865 246823
World Wide Web home page: http: //www.bios.co.uk/

Published in the United States of America, its dependent territories and Canada by Springer-Verlag
New York Inc., 175 Fifth Avenue, New York, NY 10010–7858, in association with BIOS Scientific
Publishers Ltd.

Production Editor: Fran Kingston
Typeset by J&L Composition Ltd, Filey, UK
Printed by The Bath Press, Bath, UK

Contents

CHAPTER **6**   **Preparation of sectioned, whole mount or centrifuged material**   72

CHAPTER **7**   **Stringency and kinetics**   90

CHAPTER **8**   **DNA:DNA *in situ* hybridization**   96

Preface

During the 1990s, *in situ* hybridization has become established as an essential method in cell and molecular biology. It is able to link DNA sequences with their organization and physical position, and RNA with its pattern of expression in tissues and during development, defining differences in gene behavior and genome evolution. Now, every issue of cell and molecular biology journals presents *in situ* hybridization results, complementing other methods of gene analysis. *In situ* hybridization answers many high priority questions as we move from study of DNA at the sequence level to the functional level, and from the whole organism to understanding development at the cellular and genetic level – functional genomics.

The rate of technological development in the field of *in situ* hybridization has been rapid: radioactive probes are now rarely used, while labeling methods, fluorochromes, chromosome and tissue preparation methods, microscopes and imaging technology have all improved enormously. However, selection and use of techniques from the literature is difficult: many developments have been evolutionary rather than revolutionary, with a plethora of different variations. Many edited technique volumes include one or more chapters on *in situ* hybridization, but such chapters rarely present a coherent and comprehensive view of the methods. Therefore, we felt there was a need for a single volume covering the key methods in molecular cytogenetics, with information about laboratory organization and costs not found elsewhere, covering the range of research applications from plants to animals.

In this volume, we have aimed to distil the key techniques to an organized, uniformly presented, efficient and effective series of protocols with enough suggestions and hints that a new user in either a molecular biology or cytogenetics laboratory can get them to work. We recognize that many variations and modifications beyond those presented may be helpful, but the methods presented, which have been transferred between many laboratories, will produce reliable, publication-quality results. The book is aimed at the practical researcher, but we hope it will be useful for other purposes, including teaching, interpretation of published results, and grant applications.

We are particularly grateful to Fran Kingston, Lisa Mansell and Jonathan Ray at BIOS for their great patience and encouragement with the preparation and production of the book, and the current and past members of our laboratories who have been instrumental in the development of the methods and tolerant during the writing stages!

Trude Schwarzacher and Pat Heslop-Harrison

Acknowledgements

The DNA *in situ* hybridization protocols are based on those developed and used over the last 10 years in the Karyobiology Group at the John Innes Centre, Norwich. We are grateful to the more than 30 laboratory members over that time who contributed to the protocols and research. We particularly thank Andrew Leitch who coordinated and wrote most of the first book of *in situ* protocols (Leitch *et al.* 1994) with Ilia Leitch and David Jackson who was responsible for the first set of RNA protocols, Gill Harrison for her continuing technical expertise and contributions, Tony Warford for his contributions to the RNA protocols, Irena Solovei and Thomas Cremer for assistance with whole mount protocols, and Peter Lichter for allowing the use of his EMBO course. Thomas Schmidt, Kesara Anamthawat-Jónsson, A. (Xana) Castilho, Anette Kamm, Nuno Neves, Andreas Katsiotis and Alexander Vershinin made special contributions to DNA hybridization techniques. We thank John Doonan, Ali Bevan and Trish Lunness for assistance with the RNA protocols, and Brian Murray and Lindsay Powles for many helpful comments.

We are particularly grateful to those who provided the illustrations: Robert Doudrick (USDA Forest Service, North Carolina), Tella Galasso (Germplasm Resource Institute, Bari, Italy), Thomas Schmidt (University of Kiel, Germany), Angeles Cuadrado (Computense University, Spain), Marian Orgaard (Royal Veterinary and Agricultural University, Denmark), Nuno Neves (Lisbon, Portugal), Julian Osuji (University of Port Harcourt, Nigeria), Melanie Gatt and Brian Murray (University of Auckland, New Zealand), Chris Gilles (University of Sydney, Australia), Helen Vaughan (University of East Anglia, UK), Peter Lichter (German Cancer Research Institute, Heidelberg, Germany), Thomas Cremer and Irene Solovei (University of Munich, Germany), the Department of Histopathology/Cytopathology and Dr RY Ball (Norfolk and Norwich Hospital, UK) and Pascale Harvey (University of East Anglia, UK), Tony Warford (Cambridge Antibody Technology), and Sybille Kubis, John Doonan and Patricia Lunness (John Innes Centre, Norwich, UK).

Disclaimer

The information in this book was obtained by BIOS Scientific Publishers Ltd from sources believed to be accurate. Attention to safety aspects is an integral part of all laboratory procedures and it remains the responsibility of the reader to make sure that the procedures that are followed are carried out in a safe manner and that all necessary national safety requirements have been looked-up and implemented. Any specific safety instructions relating to items of laboratory equipment or particular products must be followed.

CHAPTER 1

The *in situ* hybridization experiment

1.1 Molecular cytogenetics and the objectives of *in situ* hybridization

Molecular cytogenetics and the methods of *in situ* hybridization have revolutionized our understanding of the structure, function, organization, and evolution of genes and the genome. They have enabled the linkage of molecular data about DNA sequence with chromosomal and expression information at the tissue, cellular and sub-cellular level and they have changed the way we apply cytogenetics to both medicine and agriculture. DNA:DNA *in situ* hybridization allows the location and abundance of a DNA sequence to be determined by hybridization of a labeled DNA probe to a DNA target in a microscope preparation – chromosomes, nuclei or DNA fibers. It can be used for physical mapping of sequences to their chromosomal location (including ordering of probes and checking for chimeric clones), the identification and characterization of chromosomes, or chromosome segments, and to provide indicators of recent or evolutionary rearrangements in the genome. We can examine and understand genome organization in species with large and small genomes and investigate the spatial distribution of DNA sequences at somatic metaphase, interphase and meiosis. *In situ* hybridization shows chromosome rearrangements and changes in sequence abundance during evolution, organism development and disease. RNA *in situ* hybridization, a parallel technique, localizes RNA and answers questions about gene expression patterns in tissue and RNA abundance – critical aspects of developmental biology, functional genomics and neurobiology. Both DNA and RNA *in situ* hybridization can detect the presence and expression of pathogen – particularly viral or bacterial – sequences in eukaryotes.

In situ hybridization is a widely used and mature technique, with many hundreds of laboratories publishing using the methods. Skills are required in a combination of cytogenetics, molecular biology, immunocytochemistry, microscopy and image analysis. This book is aimed at those who wish to learn and use efficient and reliable protocols for *in situ* hybridization to both DNA and RNA targets. It is also aimed at those researchers who wish to check the validity and interpretation of published *in situ* hybridization data, and to decide when *in situ* hybridization is necessary for their own research. Details of medical applications of *in situ* hybridization and specialized techniques currently used mostly for human cytogenetic analysis are discussed, but not covered by protocols. As in much modern biology, inter-laboratory collaboration is valuable because of the investment in time, materials and skills required for any advanced technique, so it is important to consider the option of carrying out projects involving some molecular cytogenetics jointly between laboratories. For mapping of a human and mouse gene probe to a chromosome region, and for gene expression studies in mouse embryos, DNA and RNA *in situ* hybridization is available as a commercial service (e.g. Genome Systems).

We give protocols required for publication-standard DNA or RNA *in situ* hybridization, with enough background information to enable trouble-shooting and data assessment; practical notes or alternatives are added to most protocols. Names and abbreviations of chemicals and buffers (Appendix 3) are widely used in laboratories; normally we have followed terminology in the Sigma catalogue or in Sambrook *et al.* (1989). Many notes are necessarily personal since they are based on our experience and queries received in our laboratory; a few repetitions are included to limit the need for cross-referencing.

Methods are given with times, concentrations and alternatives which we know work successfully, although we recognize that the presentation makes some points overly dogmatic or prescriptive. Many variations or alternatives to protocols, not mentioned here, are also successful, and improvements can be made to many steps with experience.

Probes and preparations are covered in detail. A separate chapter describes the theoretical basis of hybridization stringency, since this important point is often misunderstood. Hybridization and detection protocols and variations are given before sections on microscopy and trouble-shooting. The final sections include frequently asked questions and references. Protocols widely used in molecular biology laboratories, including those for DNA extraction, gel electrophoresis and DNA cloning, are beyond the scope of the book. These are available in most research departments, and they are well covered in numerous books, molecular biology catalogues and Internet sites. We particularly recommend Sambrook *et al.* (1989) which, despite much competition and rapid advances in molecular biology, remains a clear, comprehensive and accurate source of experimental protocols. Quality control of all DNA coming into the molecular cytogenetic laboratory is essential: in our own laboratories, receipt of incorrect or poor quality material from outside has cost many experiments. Essential protocols for such quality control are given here.

DNA *in situ* hybridization to chromosomal targets has become almost synonymous with fluorescent *in situ* hybridization (FISH). Compared to other methods, fluorescent detection of hybridization sites has advantages in resolution, contrast, speed and safety, as well as the critical advantage of simultaneous multi-target localization. Radioactive methods, used for the first *in situ* hybridization experiments to chromosomes as long ago as 1969 (Gall and Pardue 1969; John *et al.* 1969), detected hybridization by photographic emulsion over the surface of the preparation, and have been used extensively for mapping of genes (Ferguson-Smith 1991). Colorimetric methods, detecting the sites of hybridization by colored precipitates with enzyme substrates, have lower spatial resolution (<1 µm), but high-cost fluorescence microscopes are not required and reagents are more temperature-and light-stable (although some are toxic). Radioactive and colorimetric methods are now largely superseded by non-radioactive, fluorescence methods covered in this book; protocols are similar up to the detection of hybridization sites.

RNA *in situ* hybridization uses radioactive, colorimetric or fluorescent detection. All are current, although radioactive methods are declining because of the difficulty of using photographic emulsions, safety, and the long exposure times needed. Fluorescent detection has limited applications as many fixed tissues and embedding media are autofluorescent. Colorimetric detection is most widely used, and gives high sensitivity: a few transcripts or pathogens can be detected at a cellular level of resolution (<10 µm).

Examples of different uses of *in situ* hybridization are shown in *Figures 1.1–1.18*.

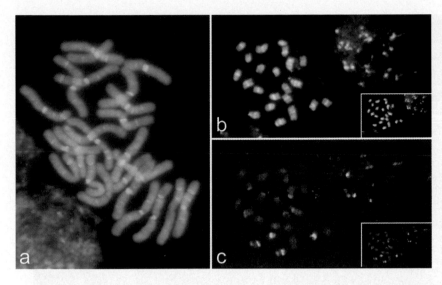

Figure 1.1 DNA *in situ* hybridization of rDNA probes to chromosomal preparations from species with contrasting genome sizes. (**a**) A metaphase from slash pine (*Pinus elliottii*; 2n = 24), a species with a very large genome. The chromosomes show DAPI bands (bright blue) and clusters of the 45S rDNA genes at 16 hybridization sites (red; see Doudrick *et al.* 1995). Separate exposures were made for red and blue fluorescent images. The negatives were then scanned, and overlaid using Adobe Photoshop. Magnification: 900×. (**b, c**) A metaphase from *Vigna unguiculata* (2n = 22) stained blue with DAPI (**b**). Double-target *in situ* hybridization, using a single exposure with a triple-bandpass filter set, shows that five pairs of chromosomes carry major sites of 18S-5.8S-26S rRNA genes (red), while two pairs carry 5S rRNA genes (green-yellow color; see Galasso *et al.* 1997). Insets: The metaphase showing the size of *Vigna* (500 Mbp genome size) in relation to pine (23 000 Mbp, in (a)) chromosomes. Magnification: 2400×.

Figure 1.2 The genomic locations of various repetitive DNA sequences on sugar beet (*Beta vulgaris*; 2n = 18) metaphases stained with DAPI (blue). (**a**) *In situ* hybridization of a *copia*-like retrotransposon (red signal) shows that the sequence is abundant and dispersed throughout the genome with a few sites of clustering. (**b**) *In situ* hybridization of a non-viral, LINE-like retrotransposon (green signal) shows a more clustered distribution than the *copia*-like elements in (**a**). (**c**) A labeled synthetic oligomere of the telomere sequence (red signal), (TTTAGGG)$_6$, hybridizes to the ends of all chromosome arms. (**d**) The species has a single major pair of 45S rDNA sites (yellow signals), and intercalary sites of variable sizes on each chromosome arm of a repetitive DNA sequence (red signals). Images courtesy of Thomas Schmidt. See Schmidt *et al.* (1995), Schmidt and Heslop-Harrison (1998) and Kubis *et al.* (1998) for further details. Magnification: 1300×.

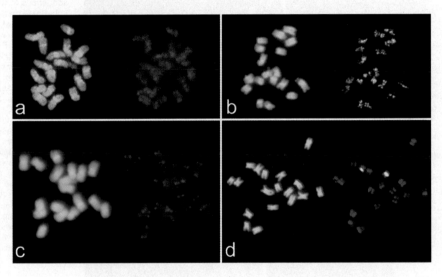

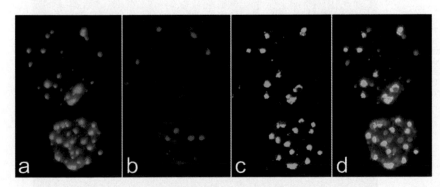

Figure 1.3 Analysis of the number of sites of two repetitive sequences isolated from *Brassica* species in two interphase nuclei of the diploid species *B. campestris* (Chinese cabbage). (**a–d**) Simultaneous hybridization with clone pBcKB4 from *B. campestris*, labeled with digoxigenin and detected with FITC anti-digoxigenin (green signal (**a**)), and clone pBoKB1 from *B. oleracea*, labeled with biotin and detected with Texas Red avidin (red signal (**b**)). In (**c**) the counterstaining of chromatin with DAPI (blue signal) and in (**d**) a computer enhanced overlay of (**a**), (**b**) and (**c**) are shown. There are about 16 green sites of pBcKB4 (some overlapping), and four sites homologous to pBoKB1 which collocalize with DAPI chromocentres and correspond to para-centromeric regions (see Harrison and Heslop-Harrison 1995). Magnification: 1300×.

Figure 1.4 Localization of a major 180 bp repetitive sequence present around all five pairs of the centromeres of *Arabidopsis thaliana* (2n = 10, genome size about 150 Mbp). (**a**) The cloned sequence shows approximately equal strength of hybridization (red) to all five chromosome pairs (counterstained blue with DAPI). (**b–e**) *In situ* primer extension (PRINS) using short oligonucleotides to variants of the AtCon sequence on metaphase (**b–d**) and pachytene nuclei (**e**) shows different signals on each chromosome pair, indicating that the variants differ in abundance. Here, the strongest signals are seen on a large metacentric (**b**), large acrocentric (**c**) and small acrocentric (**d**) chromosome pair, while one pachytene chromosome shows a strong and elongated signal (**e**). Further information about the sequence is given in Heslop-Harrison *et al.* (1999). Magnification: 5000×.

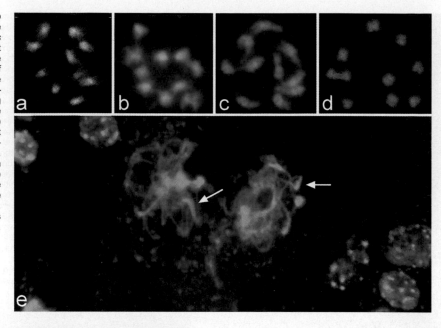

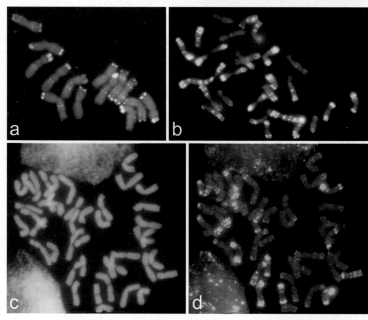

Figure 1.5 *In situ* hybridization of repetitive DNA sequences to metaphases of wild and cultivated cereals. (**a**) A metaphase of rye (2n = 2x = 14) showing hybridization of pSc119.2 to all chromosomes (yellow of the FITC) with propidium iodide counterstaining (red-orange). Because of the counterstaining, some minor intercalary sites are obscured (see *Figure 1.6*). (**b**) A metaphase of the wild wheat *Aegilops ventricosa* (2n = 4x = 28) visualized using a triple band pass filter, showing blue DAPI staining of all chromosomes and hybridization of dpTa1 (yellow-green of FITC) to most chromosomes, although the sequence is much more abundant on chromosomes of one of the ancestral genomes (the D genome); the 45S rDNA sites are labeled red (Cy3), and one major pair and multiple minor sites can be seen. Details of the karyotype are in Bardsley *et al.* (1999). (**c–d**) A metaphase of bread wheat (2n = 6x = 42) labeled with a simple sequence repeat (red) and dpTa1 (green). As in (b), dpTa1 hybridizes strongly to the D genome chromosomes, while the SSR is located in broad centromeric regions, particularly on the B genome chromosomes. Details of SSR banding patterns are given by Cuadrado and Schwarzacher (1999). Magnification: 750×.

Figure 1.6 A single metaphase of rye probed with three repetitive sequences, with the chromosomes cut out and aligned. The top row shows the cyan DAPI stained chromosomes, with all hybridization signal overlayed underneath. The lower three rows show the different chromosomal distributions of three repetitive sequences; the metaphase was probed first with pSc200 (labeled with biotin and detected with Cy3, red) and pSc250 (labeled with digoxigenin and detected with FITC, green), and then reprobed after photography with pSc119.2 (labeled with biotin detected with Cy3, displayed in blue). Details of clones are described in Vershinin *et al.* (1995). Magnification: 750×.

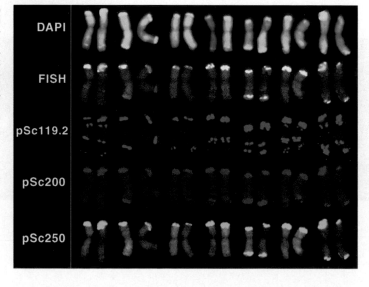

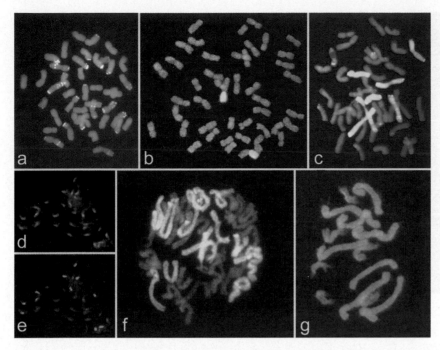

Figure 1.7 Examples of genomic *in situ* hybridization. (**a**) A metaphase of a bread wheat (*Triticum aestivum*, 2n = 6x = 42) breeding line with two chromosomes substituted by chromosomes from the wild diploid wheat species *Aegilops umbellulata*. Double-target hybridization of total genomic DNA from *Ae. umbellulata* (detected in red) and a repetitive sequence pSc119.2, labels the two alien chromosomes of *Ae. umbellulata*. Details of different lines derived from wheat and *Ae. umbellulata* are described in Castilho *et al.* (1996). (**b**) Another wheat breeding line (2n = 6x + 2 = 44) with a pair of chromosome arms from rye (labeled green with genomic DNA from rye) translocated onto a wheat chromosome (1B–1R) and an additional chromosome pair from the wild species *Thinopyrum bessarabicum* (labeled red with genomic *T. bessarabicum* DNA). There is some cross-hybridization of the red probe to the smaller wheat chromosomes, belonging to the D genome (see Schwarzacher 1996). (**c**) A metaphase of a hexaploid (2n = 6x = 42) hybrid involving four different genomes: the A and B genomes of wheat (showing only the blue DAPI counterstaining), rye (seven large chromosomes labeled green with genomic DNA from rye), and a wild barley species, *Hordeum chilense*, labeled red with a *Hordeum*-specific tandemly repeated sequence (see Lima-Brito *et al.* 1997). (**d, e**) A metaphase of a triploid plantain banana cultivar (2n = 3x = 33). Labeled genomic DNA *Musa balbisiana* (B genome) hybridizes to the centromeric regions of the 11 chromosomes originating from the B genome (e, red) (see Osuji *et al.* 1997). (**f**) A hybrid between *Hordeum procerum* and *Leymus racemosus* hybridized with labeled genomic DNA from *H. procerum* (green) and *L. racemosus* (red) (see Schwarzacher 1996). (**g**) A metaphase from a *Crocus* cultivar, 'Golden Yellow' following *in situ* hybridization of labeled genomic DNA from *C. flavus* (2n = 8, labeled red) and *C. angustifolius* (2n = 12, labeled green). Two sets of chromosomes were clearly distinguished, confirming the view that the cultivar is a triploid of hybrid origin (see Orgaard *et al.* 1997). Magnification: 750×.

Figure 1.8 A metaphase of the wheat × rye hybrid triticale showing combined silver staining to examine rRNA expression with *in situ* hybridization to see rRNA loci and the genomic origin of chromosomes. (**a–d**) Silver staining showing four expressed silver-stained rDNA sites (**a**, arrows). Sequential *in situ* hybridization with a 45S rDNA probe (**b**, green) and genomic rye DNA (**c**, red) shows that the two major rDNA sites of rye origin (**d**, on chromosome 1R, green *in situ* signal of rDNA on red, rye-origin, chromosomes) are inactive. Courtesy of Nuno Neves. See Neves *et al.* (1995b) for details. Magnification: 750×.

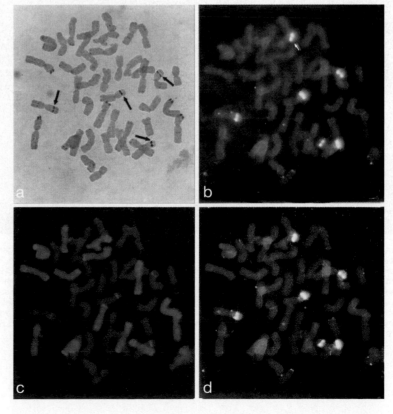

Figure 1.9 Demonstration of pararetro-virus introgression into the genome of banana. (**a–d**) *In situ* hybridization of double-stranded banana streak virus (BSV) DNA to DAPI-stained chromosomes (**a, c**) showing presence of homologous sequences at a number of red sites (arrows, **b, d**). Triploid plantain (**a, b**) with two large and some smaller sites; dessert banana (**b, d**) with eight sites (both 2n = 3x = 33) (see Harper *et al.* 1998, 1999). (**e–g**) Hybridization to extended DNA fibers. Multiple DNA fibers spreading out from nuclei of banana (**e**), labeled with a probe for BSV (yellow) and an associated sequence (red). A microscope field (**f**) showing a single DNA fiber from a banana nucleus spread to its full molecular length (see Osuji *et al.*, 1999, for details of this experiment); the row of green hybridization sites can be seen, along with randomly placed background dots. For analysis and comparison (**g**), multiple individual fibers such as those in (**f**) are aligned and oriented before measurement and making a consensus. Here, two groups of three similar fibers showing the localization of two different probes (red and green) have been aligned. Magnification: 1000×.

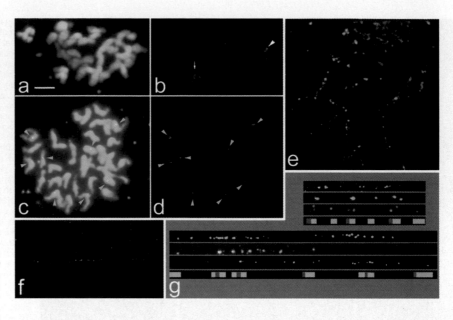

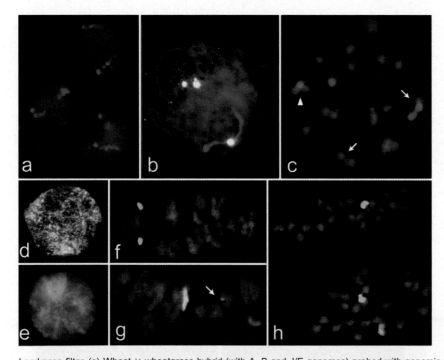

Figure 1.10 Chromosome organization at interphase and meiosis. (**a, b**) A wheat variety carrying a 1B-1R translocation following genomic *in situ* hybridization with DNA from rye (red, directly labeled with rhodamine) shows the two rye chromosome arms in separate domains at interphase (**a**) and paired at pachytene of meiotic prophase I (**b**). In (**b**), the 45S rDNA sites are visible as yellow fluorescence (directly FITC labeled), two of wheat origin and one of rye origin. (**c**) An interspecific hybrid between *Dahlia dissecta* (red with HNPP fluorescence detecting digoxigenin labeled genomic probe) and *D. merckii* (blue with DAPI counterstain) at metaphase I of meiosis showing many univalents, but some bivalents and trivalents (arrows) and a tetravalent (arrowhead) between red and blue chromosomes (see Gatt *et al.* 1999; micrograph courtesy of Melanie Gatt and Brian Murray, University of Auckland). (**d, e**) Interphase nuclei of hybrid lines after genomic *in situ* hybridization showing that the chromosomes of different genomes are organized in distinct domains. In (**d**), *Hordeum procerum* (labeled with biotin and detected with Texas Red avidin) × *Leymus racemosus* (labeled with digoxigenin and detected with FITC anti-digoxigenin) is viewed with a double band pass filter. (**e**) Wheat × wheatgrass hybrid (with A, B and J/E genomes) probed with genomic wheatgrass (*Lophopyrum elongatum*) DNA labeled with biotin, detected red, and *Triticum monococcum* DNA labeled with digoxigenin, detected green. Multiple radial sectors are seen corresponding to domains occupied by A genome (brighter yellow), B genome (brown) and J/E genome (red-brown) chromosomes (see Kosina and Heslop-Harrison 1996). (**f–h**) A wheat breeding line (2n = 6x + 1 = 43) with a pair of chromosome arms from rye (green with direct FITC) translocated onto a wheat chromosome (1B–1R) and an additional chromosome from the wild species *Thinopyrum bessarabicum* (orange-red with direct rhodamine). At metaphase I of meiosis (**f, g**), the translocated wheat-rye chromosome pair forms rod (**f**) or ring (**g**) bivalents, and segregates normally at anaphase I (**h**). The single *T. bessarabicum* chromosome forms a univalent (with a possible *de novo* translocation in (**g**) onto a wheat chromosome (arrow)) and shows nondisjunction at anaphase I (**h**) (see Schwarzacher and Heslop-Harrison 1995). Magnification: (**a, b, d–h**) 700×; (**c**) 1400×.

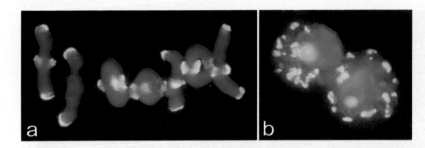

Figure 1.11 Rye chromosome organization. (**a**, **b**) *In situ* hybridization with two repetitive sequences detected with CY3 (red) and FITC (green, but seen as yellow where green and red overlap) and DAPI staining (blue), visualized simultaneously by a triple band pass filter, shows subtelomeric locations. At metaphase I (**a**) chiasmata occur close to the telomeres that show clustering at the periphery of somatic interphase nuclei (**b**). Image courtesy of Chris Gillies. Magnification: 1000×.

Figure 1.12 Chromosomal organization of simple sequence repeats (SSRs) in wheat and rye. (**a**–**d**) Locations of the SSR probes GACA (**a**) and AAC (**b**, **d**) (red), and tandemly repeated satellite DNA sequence (green). Chromosomes were counterstained with DAPI (blue in **a** and **b**). In wheat metaphases (**a**), SSRs are located in the heterochromatic N-bands. In rye metaphases (**b**), SSRs are clustered at intercalary positions giving a diagnostic distribution pattern (see Cuadrado and Schwarzacher 1998). SSRs are generally not found in the subtelomeric heterochromatin that contains tandemly repeated DNA families. Surface spreading of anther meiocytes at pachytene, reveals the synaptonemal complex after staining with silver nitrate (shown in white in **c**, **d**) and shows the contrasting organization of satellite repeats in tight clusters (**c**, arrow heads) and SSRs in extended rows of signal (**d**, arrow). Image courtesy of Angeles Cuadrado. Magnification: (**a**, **b**) 750×; (**c**, **d**) 500×.

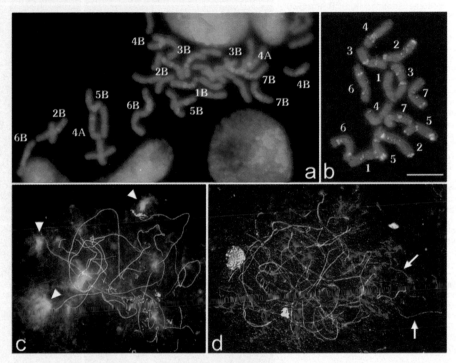

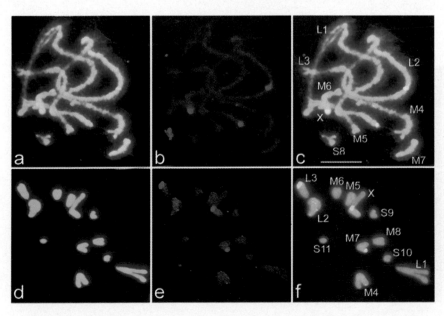

Figure 1.13 *In situ* hybridization to orthopteran grasshopper meiotic chromosomes. Localization of an 18S-25S ribosomal DNA probe (red fluorescent label) to nuclei stained with DAPI (blue). Suggested chromosome numbers are indicated on the overlay of the two micrographs. (**a**–**c**) *Chorthippus parallelus parallelus* pachytene nucleus showing three rDNA sites. (**d**)–(**f**) *Podisma pedestris* half meiotic anaphase I showing three sites, one pericentromeric on the X chromosome and two located distally, on a large and medium autosome (see Vaughan *et al.* 1999). Magnification: 800×.

Figure 1.14 Human metaphase chromosomes probed with defined repetitive sequences, single copy clones and chromosome paints. (**a**) Simultaneous hybridization of three chromosome-specific repetitive probes: a red Cy3-labeled sequence from chromosome 1, a green FITC-labeled sequence from chromosome 12 and a Cy5-labeled sequence from chromosome 5, shown in orange after processing. Chromosomes in the normal metaphase are counterstained with DAPI. (**b**) 'Reverse painting' of a normal chromosome spread (counterstained blue with DAPI) with PCR-amplified and labeled DNA from an abnormal flow-sorted derivative chromosome 1, which labels parts of both chromosome 1 and 14 in green. (**c**) Reverse painting of normal chromosome spreads with PCR-amplified and labeled DNA from an abnormal flow-sorted derivative chromosome 13, seen to paint parts of chromosome 13 and chromosome 5q. (**d**) Double-target hybridization of two cosmid probes on chromosomes of the Burkitt's lymphoma cell line JARC BL 16: cosmid-IgH is on chromosome 14q (red, rhodamine detection) and cosmid-c-myc on chromosome 8q24 (green, FITC detection). (**e**) Comparative genomic hybridization (CGH) using genomic DNA of cell line Colo320 HSR (homogeneously staining region), detected in green, with normal (reference) DNA detected in red, probed to a normal metaphase without counterstaining. Multiple regions are amplified in the Colo320 line (yellow chromosomal regions) with particular amplification at the sites with arrows. (**f**) CGH with genomic DNA of cell line HL60 (detected in green) with reference DNA (detected red) on normal chromosome spreads counterstained with DAPI. Sites of amplification and deletion in the HL60 line are seen as more yellow and more red regions respectively. The images were kindly generated within the 1996 EMBO course 'Non-isotopic *in situ* hybridization to metaphase chromosomes and inter-pahse nuclei' and kindly provided by Peter Lichter, DKFZ, Heidelberg. Magnification: 1000×.

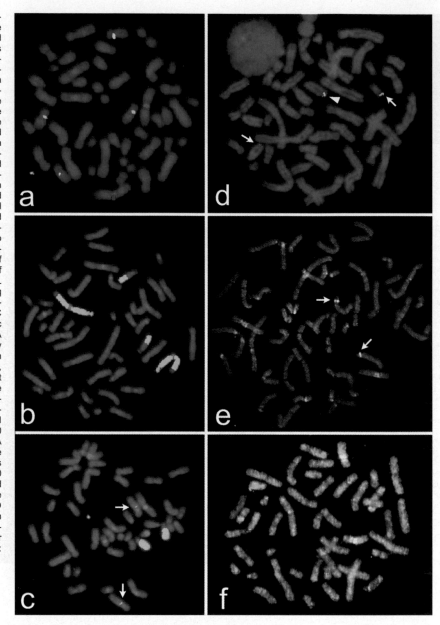

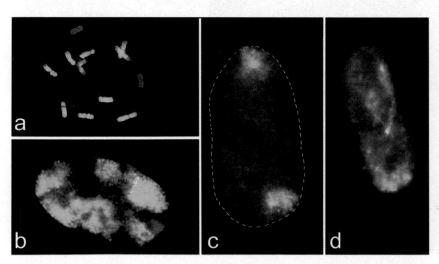

Figure 1.15 Three-dimensional organization of human nuclei analyzed by confocal optical sectioning. (**a**) A female human metaphase (2n = 46, XX) with the X chromosomes painted red and chromosomes 1, 2, 3, 4 and 5 all painted green using specific *in situ* hybridization probes. The remaining 34 chromosomes are not visible. (**b**) An interphase nucleus with the same probes as (**a**), showing the X chromosomes (red) lying in specific domains. (**c**) A human interphase nucleus (dotted outline) labeled with paints for chromosome arm 1p (red) and 1q (green), showing the domains for individual arms. (**d**) Multiple chromosome paints (red, blue, yellow, green labels) in a nucleus showing the domains occupied by individual chromosomes. Images courtesy of Thomas Cremer and Irene Solovei, University of Munich. Magnification: 1000×.

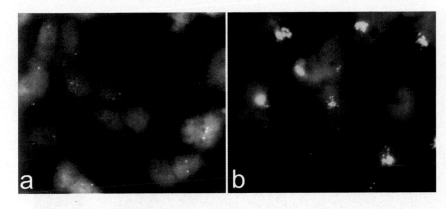

Figure 1.16 *In situ* hybridization of the oncogene probe HER-2/*neu* (Oncor/Ventana) labeled with FITC (green) to benign epithelial cells (control, **a**) and cancer cells (**b**) on the same microscope preparation; nuclei are counterstained with propidium iodide (red). Two points of hybridization are seen in the benign cells, but the gene is strongly amplified in the cancer cells (large area of green-yellow hybridization sites). The tumor was a grade 3 invasive ductal carcinoma of the breast, biologically an advanced malignancy, as 5 of the 11 lymph nodes contained metastases. The tumor material was taken from archival paraffin wax blocks from the Dept of Histopathology/Cytopathology, Norfolk and Norwich Hospital; we thank Pascale Harvey, School of Biological Sciences, University of East Anglia and RY Ball, Norfolk and Norwich Hospital.

Figure 1.17 *In situ* hybridization to RNA and DNA targets in formalin-fixed paraffin wax section preparations of human tissues. (**a**) Human cytomegalvirus DNA demonstrated in an ulcerated bowel using a recombinant DNA probe labeled with digoxigenin coupled with NBT/BCIP alkaline phosphatase visualization. (**b**) Proglucagon mRNA demonstrated in a human islet of Langerhan using an oligonucleotide probe cocktail labeled with fluorescein coupled with NBT/BCIP alkaline phosphatase visualization. (**c**) Histone mRNA demonstrating cells in S-phase in human tonsil using an oligonucleotide probe cocktail labeled with fluorescein coupled with NBT/BCIP alkaline phosphatase visualization. (**d**) *In situ* hybridization (ISH) followed by immunocytochemistry (ICC): ISH shows latent Epstein–Barr infection in Hodgkin's disease using a digoxigenin-labeled oligonucleotide cocktail directed against Epstein–Barr early RNA coupled with NBT/BCIP alkaline phosphatase visualization (more blue), while ICC labels T cells using a polyclonal anti-CD3 coupled with diaminobenzidene peroxidase visualization (brown). Images kindly provided by Tony Warford, Cambridge Antibody Technology.

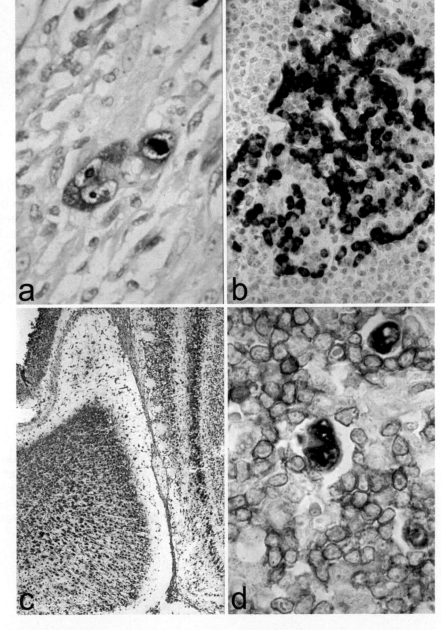

Figure 1.18 *In situ* hybridization to RNA targets to examine patterns of gene expression in apical meristems of *Antrirrhinum*. Consecutive sections of a floral meristem hybridized with anti-sense RNA probes (hybridizing to the sense RNA within individual cells) for (**a**) *cyclin1*, (**b**) *cdc2d* and (**c**) *cdc2c*; each shows a unique distribution by precipitation of enzyme substrate, while cell walls are seen by calcofluor white fluorescence staining (light blue, with addition of yellow through light illumination). Double labeling of inflorescence apicies of *Antirrhinum* with a fluorescein (red) and digoxigenin (blue-black) labeled probes viewed by transmitted light microscopy with filters to maximize contrast and differentiation: (**d**) after detection of only histone H4 RNA in red; (**e**) after detection of both histone H4 and *cdc2d* (blue-black) RNA; and (**f**) processed overlay of (d) and (e) to show differential expression of the two genes in different cells. Images kindly provided by Patricia Lunness and John Doonan, John Innes Centre; see Fobert *et al.* (1996) for further details.

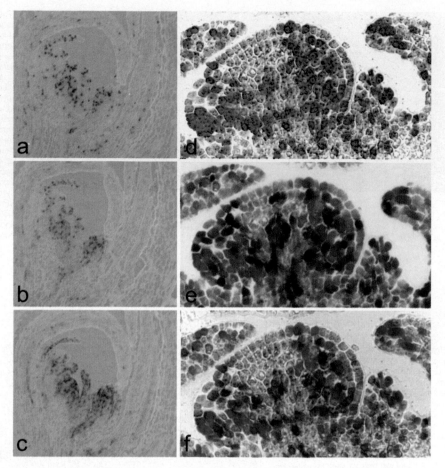

1.2 Methods and developments

The primary journals include the most recent work on the subject, and technical breakthroughs are normally reported almost simultaneously in journals and at conferences. Increasingly, even academic work is being covered by patents, so may be published in that form first, and may be difficult to find. Other developments may remain as 'trade secrets' – for example in labels, buffers or detection reagents – and may not be available other than from manufacturers.

Company catalogs and technical notes contain exceptionally valuable data of all types, and catalogs from several suppliers should always be to hand. Reagent manufacturers are a useful advice source, and the Internet provides much useful information (Chapter 13).

1.3 The design of the book

The flow chart in *Figure 1.19* shows the major division of chapters.

1.4 Timings

In a project involving extensive *in situ* hybridization, typically one person will carry out an *in situ* hybridization experiment to eight target slides every 2 weeks. Faster rates may be possible if probes are pre-prepared and less analysis is required, and the time taken can be much longer if probe isolation and characterization, or material preparation, are slow. For the protocols presented in this book, typical timings are shown in the appropriate box.

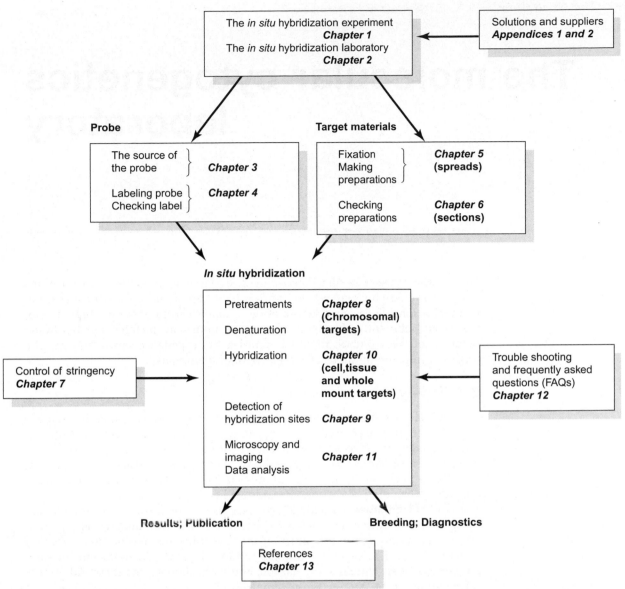

Figure 1.19 The division of chapters.

Timing for experiments

Cell pretreatment and fixation: 1 day spread over 1 week (e.g. blood culture, seed germination).
Preparation of slides and quality assessment: 2 days.
Probe isolation (plasmid mini-preparation or genomic DNA isolation): 1 day spread over 2 days.
DNA labeling and quality control: 1 day spread over 2 days.

Hybridization set up: $^{1}/_{2}$ to 1 day.
Detection: $^{1}/_{2}$ day (typically following overnight hybridization).
Preliminary observation: $^{1}/_{2}$ day.
Photography: 1 day.
Collation of photographs, re-check of problems: 1 day.
Interpretation of photographs: 1 day.

CHAPTER 2

The molecular cytogenetics laboratory

This chapter reviews the basic techniques, equipment, reagents and conditions required for carrying out *in situ* hybridization experiments. As with any other method, a suitably designed laboratory is a major help to obtaining high quality results and efficient, safe use of resources – both consumables and hardware. Many points are common to all laboratories, but experience shows they may be forgotten in planning molecular cytogenetics experiments.

2.1 Safety

All protocols presented here must be carried out safely. In most countries, there are legal obligations placed on those using and directing use of laboratory techniques. In this handbook, we have aimed to ensure protocols are accurate and safe, but it does, of course, remain the responsibility of the user to ensure they carry out procedures safely, and knowledge about hazards may change. Only researchers qualified in handling of toxic and hazardous chemicals should carry out protocols. The head of laboratory is usually directly responsible for ensuring that the staff are competent to understand the safety implications of the work they are carrying out. Manufacturers or distributors must ensure that those purchasing their products know about their safe use and disposal, and must provide relevant information; on-package labels should show hazards and users should note these. Additional information can be obtained from suppliers.

In all procedures, the use of good laboratory practice is essential – wearing of laboratory coats, using safety glasses for eye protection and face shields where appropriate, washing of hands, not eating and drinking in laboratories as a minimum. Many protocols must be carried out in fume hoods, while latex or plastic gloves are essential in some protocols both to protect the scientist from the chemicals used and to protect the reagents from contamination or degradation by substances from the scientist.

Many reagents used during *in situ* hybridization are flammable, toxic, carcinogenic or antigenic, while cultures (bacterial, viral or eukaryotic cells), plants and animals, and antibodies carry an infection and antigenic risk. The hazards associated with many fluorochromes and fluorophores are unknown but likely to be high since they bind DNA. Particular care to avoid personal exposure must be taken during weighing out; if possible, buy solutions or pre-weighed vials to avoid weighing toxic powders. Clean and tidy work is needed to avoid accumulation of dirt on the slides that can cause background signal. After the fixation stages, aseptic conditions are not normally necessary other than ensuring that solutions are not contaminated.

The ultraviolet light from UV bulbs in fluorescence microscopes and transilluminators is hazardous to eyes and skin. Skin and particularly eyes must not be exposed to illumination from the bulb during analysis of stained gels or microscope bulb alignment. A full-face mask should be worn for UV protection. Light leakage from gaps in the microscope should be covered from view with aluminum foil (without blocking cooling vents).

2.2 Reagents, solutions and equipment

Reagents and suppliers

Unless noted, reagents used in these protocols should be 'molecular biology grade' where appropriate or an equivalent high quality grade for buffer salts and acids. 'Molecular biology grade' has become a *de facto* standard, indicating that the reagent is tested for use in molecular experiments. Appendix 2 lists major suppliers of reagents for molecular cytogenetics. Prices and quality from the different suppliers tend to be similar, so quality of service and advice, often country- or region-specific, are key issues in choosing suppliers. Ask suppliers for details of protocols, published results from their reagents, names of other local users and usage tips; support the suppliers that are helpful. Kits for the 'molecular biology' aspects of *in situ* hybridization are valuable in a molecular cytogenetics laboratory, where labeling, cloning and other procedures may not be carried out daily. The manufacturers test the reagents together and provide a reliable protocol, although often the protocol can be modified to use smaller amounts of reagents. You are also paying for the technical advice that the company gives, so make use of telephone and E-mail helplines. Complain and ask for refunds where reagents do not work.

Water and making solutions

Throughout this book, 'water' refers to distilled or deionized water that is used to make up all solutions of more than 1 ml. If microbiological contamination is a problem, this can be sterilized. When solutions are made up, those that are stored at room temperature or for long periods at 4°C should be sterilized by autoclaving (simple salt solutions) or by passing through a 0.22 μm filter. Serious contamination will be seen as cloudiness, and if this occurs in recently used solutions, greater care should be taken to maintain sterility, use fresh solutions or make smaller aliquots. For RNA *in situ* hybridization, precautions against RNase activity need to be taken and are described in Chapter 6.

For smaller volume reactions and hybridization mixtures, very high quality, sterile deionized or double-distilled water (18 MΩ resistivity) kept at −20°C should be used. Because water quality is so important, many kits include water, and it is often worth purchasing 'molecular biology grade' water.

Appendix 1 lists protocols for making standard solutions and buffers that are used throughout the book.

Equipment

The box on p. 14 shows the major items of general laboratory equipment required for *in situ* hybridization.

Glassware and plasticware

Most specialist glassware and plasticware is described in protocols. Microcentrifuge tubes and pipet tips should be autoclaved (in containers of 50–100) before use to degrade organic materials and remove volatiles from the manufacturing process. For RNA *in situ* hybridization, glassware and plasticware should be new and sterile, or glassware should be baked (Chapter 6).

Laboratory equipment *for* in situ *hybridization*

−20°C freezer (Note 1)

4°C refrigerator for chemical and fixation storage (Note 2)

4°C refrigerator for cold treatment of cells and storage of photographic material and seeds

Oven at 37°C (sometimes 42°C) for hybridization incubations

Incubator/oven for cell culture or seed germination (Note 2)

Fume hood

Hot plate or gas burner

Pressure-cooker or autoclave

Set of variable-volume micropipets: 20 μl, 200 μl, 1 ml (Note 3)

Microcentrifuge (Note 4)

Vortex mixer

Dry ice or liquid nitrogen container (Note 5)

Water baths with ±0.5°C temperature control (two or more are useful, preferably one with cooling or in a cold room)

Polymerase chain reaction (PCR) machine.

Gel electrophoresis equipment for probe testing and quality control

UV transilluminator

Microwave oven

Embedding oven (whole mounts, Chapter 6)

Microtome (sections, Chapter 6)

Heated block or PCR machine with flat plate for slide denaturation

Dissecting microscope

Phase contrast microscope

Fluorescent microscope (see Chapter 11)

Microcomputer (see Chapter 11)

Notes on equipment

1. Frost-free freezers must not be used. Other domestic freezers are suitable, but note that they may contain many times their own value in reagents.

2. Fixatives and other chemicals must be kept away from living cells and photographic materials – even small concentrations of vapors will fog film, and reduce cell or seedling growth, and the metaphase index. In particular, one refrigerator and incubator should be used for chemical or fixation storage or hybridization of probes, and another for growth, storage or cold treatment of living cells.

3. Variable-volume micropipets (Gilson is the most widely used brand, and is both reliable and easy to maintain) are essential for the protocols, in at least the 20 and 200 μl sizes. Other sizes (1 μl, 10 μl and perhaps 5 ml) are useful. The rubber O-rings and plastic washer inside Gilson and many other makes of pipets need replacing at least every year, or after contamination. We emphasize the importance of accurate and repeatable pipeting of small volumes. Both the accuracy of the pipets and operator should be tested regularly with a milligram balance: 1 μl (= 1 mm^3) of water weighs 1 mg. To increase reproducibility, each operator should use only one set of pipets.

4. Centrifugation of microcentrifuge tubes (0.5, 1.5 or 2 ml) is required to bring small amounts of solutions to the bottom after mixing and to precipitate DNA. It is desirable that all lengthy centrifugation steps are carried out at 4°C or below. It is important that the centrifuge motor does not heat tubes.

5. Expanded polystyrene is safer than glass vacuum flasks which can break at such temperatures.

2.3 Laboratory techniques

The protocols presented here assume familiarity with basic molecular biology laboratory techniques. All enzymes, nucleotides and labels should be stored at −20°C. Do not keep them in a frost-free freezer. Wear latex gloves when handling enzyme tubes and use only sterile tips when dispensing. Where solutions are frozen, those containing proteins (including enzymes) should not be thawed by heating above 10°C, nor refrozen multiple times (make and freeze small aliquots). Most commercial enzymes contain glycerol and remain liquid at −20°C, so do not need thawing. Always add the enzyme as the final component of a reaction – never add the enzyme to water or concentrated buffers. Where appropriate, make up master mixes and divide for individual reactions. The working area should have a container of crushed ice, and all vials of enzymes, nucleotides, DNA and reaction mixtures should be held in the ice except during reaction incubations.

For temperature control or reactions, push microcentrifuge tubes through a hole in a thin sheet of foam or expanded polystyrene and float in a water bath. For below-ambient reactions, if cooled systems are not available, a heating water bath can be run in a cold room, or water in an expanded polystyrene ice container can be kept at 15°C with occasional addition of ice.

All manipulations of fluorochromes and fluorophores should be carried out in low lighting levels to avoid breakdown of reagents – normal levels of electric lighting are satisfactory, but bright daylight or sun coming through windows may give problems. Store the reagents in dark or foil-wrapped containers, and incubate reaction vials in darkness; slide preparations should be stored in the dark.

2.4 Laboratory layout

Efficient physical layout of the laboratory is important for achieving the highest quality and efficient throughput. It is convenient if communal facilities and consumables are arranged around the periphery of the laboratory. Safe storage for prints, negatives, and computer archives should be built into the laboratory design. DNA samples and clones should be duplicated and stored in freezers at more than one site in case of power or freezer failure.

It is critical that fluorescence microscopes are located in a totally dark and comfortable room. We emphasize the requirement for a dark, suitably equipped room, as it is frequently ignored in our experience, and makes it impossible to obtain good results from a microscope costing many tens of thousands of dollars. If a suitable room is not made available, it is impossible to visualize, analyze and photograph fluorescent *in situ* hybridization. The microscope operator will spend many hours in the room: it must have a comfortable chair where the height and back can be adjusted, must be well ventilated, with no people opening the door or walking through, and the microscope must be on a vibration-free table. A small, low power light connected to a convenient hand-controlled dimmer switch, which can be reached while sitting at the microscope, is required for taking notes and changing slides. Standing up to adjust lighting destroys concentration. It takes the eyes 10 minutes or more to become dark-adapted so *in situ* hybridization signals can be seen, and as soon as room lighting is turned on or a door opened, this adaptation is lost. Even very bright signals (often accompanied by weaker but important signals) cannot be clearly evaluated with any external light.

2.5 Costs

A single series of *in situ* hybridization experiments in a cytogenetic laboratory can be set up for a few hundred dollars, involving purchase of labeled nucleotides, labeling kits, and reagents for hybridization and detection. In a research project with extensive *in situ* hybridization, the consumable costs are substantial: in an academic research laboratory, each bench researcher will use per year up to US$12,000, UK£10,000, ¥2,000,000, or €20,000 worth of consumables and photography (using exchange rates suitable for molecular biology supplies). On top of this, the same amount of money is required every 3 years for replacement of hardware – a set of automatic pipets, water bath, microcentrifuge, domestic freezer, one filter set and an objective for the fluorescent microscope, and a basic microcomputer and software. In different terms, the cost per successful slide with double-target *in situ* hybridization, including average material preparation costs, probe labeling, hybridization and detection reagents, microscope bulb costs, and photography, but not time or overhead charges, is about US$70. For multiple labeling systems, such as those used to label all human chromosomes, the cost may rise to 10 times this.

Breakdown of consumable costs for *in situ* hybridization

Probe preparation	10%	Glassware and plasticware	10%
Probe labeling reagents and enzymes	20%	Bulbs for microscopes	5%
Specimen preparation	10%	Photography	15%
Probe testing reagents	5%	Data analysis – computer media, printing	10%
Detection reagents, antibodies, fluorochromes and mountants	15%		

Publication costs are additional to these, and journals may charge US$500–1000 or more for color reproduction and perhaps surcharge reprints. Major primary journals in this field with free color include *Annals of Botany, Development, Protoplasma, Journal of Cell Science, Virology* and (for digital plates) *Chromosoma*.

While allowable costs vary between grant agencies, a typical breakdown, based on the annual costs above, is given in the box on p. 15.

CHAPTER 3

The probe

The probe is the fragment of DNA or RNA that is labeled for detection during the *in situ* hybridization experiment. Methods for cloning, isolating and handling DNA sequences can be found in standard molecular biological textbooks (e.g. Sambrook *et al.* 1989) and are not discussed here. Old and Primrose (1992) and Allen (1999) give valuable applied and theoretical background information about molecular biology. As with other methods, kits are available for most standard procedures. Some pre-labeled DNA probes are available commercially.

Whatever the experiment, the source and nature of the DNA or RNA to be used as probe must be carefully documented. Confusion in probes is frequent (perhaps because they are stored for years in small, hard-to-label vials in freezers) and can take a long time to resolve. Clones and probes should be thoroughly checked before use, to determine the size, location of known restriction enzyme sites, lack of degradation of the nucleic acid, probe concentration and other parameters.

3.1 Types of probe

Any source of DNA or RNA can be used to obtain the probe for *in situ* hybridization depending on the requirements of the experiment. Cloned DNA sequences are perhaps used most frequently, but total genomic DNA, PCR products and synthetic oligonucleotides are also common. Often, molecular information about the sequence, copy number, genomic organization and species specificity of the probe is known, and may give the reason for carrying out the *in situ* hybridization experiment. DNA *in situ* hybridization will give information about the long-range organization, chromosomal location and genomic distribution, copy number, evolutionary changes and interspersion with other sequences, which cannot be obtained by other methods. RNA *in situ* hybridization gives information about the location and amount of expression of particular genes.

Cloned probes

Clones have a stretch of DNA from the species of interest inserted into a vector and amplified in host cells of an appropriate organism. Plasmid clones (*Figure 3.1*), where the clone is in vectors such as pUC18, pUC19 or pBluescript, and carried in *Escherichia coli* cells are used most frequently, with cloned DNA inserts typically less than 10 kb long. Larger DNA inserts may be carried in bacteriophage (e.g. lambda or M13), cosmid, or, in bacteria or yeast artificial chromosome (BAC or YAC) vectors where the cloned DNA insert may be over 1 Mb long. The DNA may come from the nuclear or cytoplasmic genomes (random or selected in some way), or from cDNA (complementary DNA) derived from mRNA by reverse-transcription. When a laboratory has published data about a clone, the authors are under an obligation to make it freely available to any researchers for academic use, and it should be sent in response to a letter; journal editors can be informed if this is not the case. Most laboratories sending out clones do not allow clones to be passed on to third laboratories.

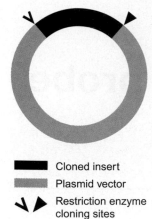

Cloned insert

Plasmid vector

Restriction enzyme cloning sites

Figure 3.1 A circular DNA plasmid with cloned insert.

Cloned DNA is extremely versatile. Inserts can be isolated in large quantities of high purity and complete fidelity to the original sequence. Double- and single-stranded DNA or RNA probes can be made from either DNA strand. The DNA can be readily labeled and used for both *in situ* and Southern hybridization experiments. For single-copy sequences, hybridization sites of DNA probes longer than 3 kb can be detected routinely, by hybridization to both chromatids on chromosomes with a target site in most metaphases. Shorter probes – less than 1 kb – can often be detected, although the signal may not always be present and background of similar strength may make interpretation more difficult. Many probes may include repetitive DNA elements and these will give a signal at many sites on chromosomes.

Genomic DNA

Total genomic DNA, isolated from the nucleus of an organism, has proved to be a versatile and informative probe (see Heslop-Harrison and Schwarzacher 1996a) for looking at the chromosomes present in hybrid cells (Pinkel *et al.* 1986; Schardin *et al.* 1985) or organisms (genomic *in situ* hybridization, GISH; Schwarzacher *et al.* 1989), chromosomal evolution and chromosome introgression (Schwarzacher *et al.* 1992a). In cancer cytogenetics, genomic DNA extracted from tumors and normal cells has been used to detect amplification and reduction in copy number by *in situ* hybridization to chromosome spreads from normal individuals (Kallioniemi *et al.* 1992; comparative genomic hybridization, CGH; Joos *et al.* 1993). Genomic DNA is also used as a template for amplification of sequences by PCR without the need for cloning.

For *in situ* hybridization, genomic DNA is also used to prevent ('block') hybridization of probe to DNA sequences in common between the probe and target. Larger DNA clones often include repetitive DNA, and it must be prevented from hybridizing over large parts of the target chromosomal DNA, while leaving the low-copy sequences in the probe available to hybridize to the chromosomes (chromosomal *in situ* suppression, CISS; Lichter *et al.* 1988; Jauch *et al.* 1990). The block is added in excess, and hybridizes both to the probe and the target. Total genomic DNA can be modified to be enriched in repetitive DNA (widely called Cot = 1 DNA from the reannealing kinetics, and available commercially for human DNA), improving the blocking of repetitive sequences in the probe while leaving low-copy sequences available for hybridization.

Flow sorting and microdissection of chromosomes

Individual chromosomes can be sorted by flow cytometry, providing a valuable way to partition the genome. Chromosomes or chromosome segments can also be microdissected, using needles or laser-based systems (Lengauer *et al.* 1991), and amplified for use as probes for *in situ* hybridization. Labeled pools of sorted chromosomes, under the hybridization conditions used, only label chromosomes homologous to their chromosome of origin, a method known as chromosome painting (Lichter *et al.* 1988; Pinkel *et al.* 1988). The method has been shown to work in many mammals and *Drosophila* (Fuchs *et al.* 1998), although not plants (Fuchs *et al.* 1996; Schwarzacher *et al.* 1997). Chromosome painting has been used extensively in studies of aneuploidy in mammals, mammalian chromosomal evolution (Scherthan *et al.* 1994; Wienberg *et al.* 1994; Yang *et al.* 1998) and detection of chromosome aberrations in cancers or radiation mutants (Arnoldus *et al.* 1991; Cremer *et al.* 1988; Hopman *et al.* 1989; Jauch *et al.* 1990; Le Beau 1993).

Chromosomes sorted from patients with genetic diseases can be used to show the sites of deletions: a pool of, typically, 200–300 abnormal chromosomes is labeled and probed onto normal chromosome spreads. The pool labels the relevant chromosome of a normal individual throughout most of its length, but leaves a gap (no hybridization) where there are no homologous sequences present (reverse chromosome painting: Carter *et al.* 1992; Joos *et al.* 1993).

PCR amplification of DNA

PCR is an efficient and fast method to amplify specific DNA sequences from clones, total genomic DNA or genomic DNA fractionated by methods such as flow sorting or microdissection of chromosomes. Detailed protocols for PCR amplification and labeling of DNA cloned in common vectors are given in the labeling chapter (*Protocol 4.3*).

Oligonucleotide primers may be used to isolate and amplify either specific or all sequences by PCR from total or partitioned (e.g. by flow sorting or microdissected) genomic DNA. Specific primers for conserved regions of sequences such as retro-transposons can be used (see additional protocols following *Protocol 4.3*). Where a DNA sequence is published, it can often be efficient to construct PCR primers at the end of the sequence and amplify the required sequence from genomic DNA (see *Protocol 4.3*) of the species originally used for isolation or other species. The DOP-PCR (degenerate oligonucleotide primed – PCR) and Alu-PCR methods have been developed to amplify multiple random fragments from large (e.g. YAC) clones or chromosomes sorted by flow cytometry or microdissection (see Telenius *et al.* 1992, additional protocols to *Protocol 4.3* and Lengauer *et al.* 1992).

Synthetic oligonucleotides

Synthetic oligonucleotides are of increasing importance for *in situ* hybridization. They can be localized, following labeling, by the normal *in situ* hybridization method where a particular motif is abundant. The sequence present at the telomeres, typically homologous to variants of TTTAGGG in plants or TTAGGG in animals (or complements, depending on the species; see, e.g., Greider and Blackburn 1989; Richards *et al.* 1993) is an example, while other simple sequence repeats may give useful hybridization patterns on chromosomes. Oligonucleotides can be localized by the primed *in situ* hybridization method (PRINS; Gosden *et al.* 1991; Koch *et al.* 1992) which, at the moment, is only reliable for repetitive DNA sequences.

It is simple, cheap (<US$0.50 per base) and fast to order oligonucleotide sequences from companies; standard (rather than the extra-cost chromatography) purification methods are adequate for *in situ* hybridization applications. The oligonucleotides are supplied as lyophilized powder, with accurately measured concentrations; purification from columns and concentration measurement are tedious to carry out in the research laboratory. Currently, oligonucleotides are usually labeled by end labeling (*Protocol 4.4*). Labels can also be incorporated during synthesis, a procedure we expect will become more economic, or by chemical modification of the oligonucleotide. In the future, we predict that *in situ* methods will enable detection of single copy oligonucleotides, at which point use of synthetic probes will become universal.

Highly repetitive DNA from restriction digests or centrifugation

The first *in situ* hybridizations were carried out with DNA fractions of a different density (satellite DNA; see Gosden *et al.* 1975), as separated by high speed centrifugation, although such studies are not now used. In many species, abundant fragments of a particular size may be detected following agarose gel electrophoresis of restriction enzyme digests of DNA. These fragments (restriction satellites) may be cloned, or labeled from the gel and used as probes for *in situ* hybridization. Relic DNA, the DNA remaining at high molecular weight after restriction enzyme digestion, is also a useful source of highly repetitive DNA sequences. Sequences collocating with heterochromatin bands (C-bands or N-bands), or with centromeric or sub-terminal locations, are often isolated by such methods.

RNA probes (riboprobes)

RNA probes are frequently used for *in situ* hybridization to RNA targets (Coen *et al.* 1990; Fobert *et al.* 1994). RNA:RNA hybrids are more stable than DNA:RNA hybrids, and background may be reduced by removing unhybridized RNA probe with ribonuclease H, an enzyme which destroys unhybridized single-stranded molecules. Normally, the sequence of interest as a probe is cloned as DNA in a vector containing transcription initiation sites for bacteriophage RNA polymerases (*Protocol 4.5*). These enzymes are then used to produce single-stranded, labeled RNA probes (riboprobes). RNA probes can be used for hybridization to DNA targets; this method is rarely used but has some advantages.

Commercially available probes

Increasing numbers of pre-labeled probes are available commercially for human chromosome applications. Both whole chromosome paints and satellite sequences to label the centromeres of all 22 human autosomes and the X and Y chromosomes are available from several suppliers (e.g. Vysis, Imagenetics, Roche, Scotlab, Life Technologies, Cambio, Sigma). DNA probes for microdeletion research, specific chromosomal regions and telomeres are sold by Oncor, while several gene probes are available such as those detecting oncogenes (e.g. *HER-2/neu*) and chromosome translocations (e.g. the Philadelphia chromosome) (e.g. from Vysis).

3.2 Preparation of DNA for use as a probe

This section is provided as a short guide to the molecular biology techniques required to obtain and control quality of DNA for labeling and *in situ* hybridization. We note particular problems found in the molecular cytogenetic laboratory, but do not repeat detailed notes and protocols available in molecular biology protocol books, in particular Sambrook *et al.* (1989).

Genomic and plasmid DNA: isolation and transport

Numerous protocols for genomic DNA isolation are available. Recent developments in kit isolation buffers and protocols (e.g. by Qiagen) give extremely efficient, high quality DNA isolation from small tissue samples of nearly any organism. Other methods described in detail in molecular biology protocol books are also effective and cheap, but may require more equipment and skill.

Plasmid DNA is most conveniently isolated using mini-preparation kits (e.g. Qiagen, Promega) from overnight cultures. These typically give a yield of 5–50 µg from a 10 ml culture of the bacteria containing the plasmid, ample for several labeling reactions. Both genomic and plasmid DNA must be free of RNA before use as probe. Many kits give DNA free of RNA, or DNA can be treated with RNase (*Protocol 3.1*). Phenol/chloroform extraction is a very effective method to purify DNA that cannot be labeled or enzyme-digested because of impurities, while cesium chloride density gradient centrifugation provides large quantities of very pure DNA. It is convenient to aim to work with DNA at a concentration of 1 µg µl^{-1} in water (for PCR amplification, typically make a further stock using a 1:100 dilution, 10 ng µl^{-1}). Although not necessary, some laboratories dissolve DNA in TE buffer (Appendix 1) which may inhibit subsequent enzyme steps, or include an excess amount of 'carrier' DNA (e.g. from salmon sperm).

DNA, including labeled probes, is best sent or carried between laboratories in solution in water (if pure and sterile), as dried (lyophilized) DNA, or in a solution to which two volumes of 96% ethanol has been added (called 'DNA in 70% ethanol'; see the first step of *Protocol 4.6*). To complete precipitation and to re-dissolve DNA in water, follow *Protocol 4.6* from the appropriate stage to obtain the DNA dissolved in water. Glycerol stocks or stab cultures of bacteria with plasmid may also be transported (and are also appropriate for long-term storage), but are

Form 3.1: Cloned DNA Data Sheet

Name of clone: Date of dispatch:

Supplied by: Address:

Telephone/FAX number: E-mail:

Original citation/reference:

Related citations:

THE MATERIAL SENT:

Form: **Plasmid**   **Bacteria**
          Isolated from: mini-prep   Glycerol stock/Stab
          PCR product   **Purified insert**

Purification method:

Lyophilized original volume: /in water/in TE/precipitated in 70% ethanol:

Amount (μg):

Date of preparation and any person or batch code:

THE INSERT:

Source organism for insert:

DNA/RNA source:

Original isolation details (full/partial enzyme digest etc.):

Length of insert:

Sequenced: Yes/No Sequence database accession no:

Can the insert be amplified by PCR?: Yes/No/Not tested/Not applicable

Primers recommended for PCR amplification:

THE VECTOR:

Cloning vector:

Cloning site:

Enzymes recommended to release the insert:

THE HOST:

Bacterial/other strain:

Selective conditions:

ADDITIONAL INFORMATION: (e.g. restriction maps, internal enzyme sites, recommended labeling protocols, other contacts)
This clone is supplied for academic use only and should not be distributed to other laboratories without permission. All publications using the clone should cite the original reference given above.

more sensitive to adverse conditions and can lose the plasmid or insert; normally they should be sub-cultured immediately on receipt. When sending or receiving DNA, ensure it is accompanied by complete data including the type of DNA and amount sent, the vector and restriction sites and any acknowledgments to the original people who isolated it, or publications that are relevant. *Form 3.1* gives the typical information required.

Carefully examine DNA before use in reactions. Unlabeled dried DNA should be a pale cream color, and after dissolving, the solution should be clear. A brown color indicates impurities that may prevent labeling. With experience, touching a micro-pipet tip to the surface of the solution enables estimation of DNA concentration: a 1 µg µl^{-1} solution will be slightly viscous and form a short string as the tip is pulled away from the surface. If the DNA is obviously impure, or any difficulties are encountered later with restriction enzyme digestion or labeling of DNA, we recommend cleaning using suitable kits (e.g. Stratagene, Qiagen, Promega) or phenol/chloroform extraction.

Ribonuclease (RNase) treatment of DNA

DNA for labeling should be free of RNA. *Figure 3.2* (Section 3.3) shows gel electrophoresis results from DNA contaminated with RNA. To remove RNA, follow *Protocol 3.1* after total genomic and plasmid DNA extraction.

3.1 | **PROTOCOL 3.1** RNASE TREATMENT OF DNA

PROTOCOL

Reagents

Ribonuclease A (RNase A), DNase-free.
Stock solution 10 mg ml^{-1} (approximately 1000 units ml^{-1}) (Note).

Method

1. Add 1/10 volume of RNase stock to RNA-containing DNA sample.
2. Incubate at 37°C for 1 h.
3. DNA can be purified by precipitation with ethanol (*Protocol 4.6*) or commercial purification kits, but most DNA labeling protocols work in the presence of RNase.
4. Removal of RNA can be checked by gel electrophoresis, but the protocol is robust so routine testing is not necessary.

Note

RNase is a very robust enzyme and will work in most buffers or solutions at temperatures up to boiling. Formerly, DNase-free RNase was prepared by boiling crude RNase preparations.

Fragmenting of large DNA molecules

Genomic DNA and DNA from large clones such as YACs or BACs may be more efficiently labeled by nick translation if it is broken into fragments of 500–1000 bp (sometimes known as 'sheared DNA'). Shorter fragments of genomic DNA, 50 and 300 bp, may be required to block hybridization of repetitive DNA sequences present in the probe or common between two species for genomic *in situ* hybridization. Several methods are available for fragmentation: autoclaving, sonication, digestion with DNase or pulling through a fine needle at the end of a syringe. *Figure 3.2* (Section 3.3) shows gel electrophoresis results from DNA before and after sonication.

Autoclaving: We use this method most often and adjust the time of autoclaving to give the fragment size required. Follow *Protocol 3.2*.

PROTOCOL 3.2 FRAGMENTING OF GENOMIC DNA BY AUTOCLAVING

3.2

PROTOCOL

Equipment

Domestic-type pressure-cooker or laboratory autoclave with suitable regulation of time and temperature.

Method

1. Use 5–50 μg DNA (less for labeling, more for use as block) in 100 μl water in a microcentrifuge tube securely closed with a clip or autoclave tape.
2. Bring up to temperature and simmer under pressure for 2 min (nick translation) to 5 min (blocking DNA) (Note).
3. For nick translation, allow to cool slowly in the pressure-cooker on the bench (to re-anneal DNA). For use as blocking DNA, cool pressure-cooker under running water, release excess pressure, remove DNA and place on ice immediately to limit re-annealing.
4. Load a fraction of the DNA onto an agarose gel for size separation (*Protocol 3.3*).
5. Check range of fragment sizes obtained; if outside the desired range of 100–1000 bp, adjust time of pressure-cooking.

Note

Time needs to be adjusted to sample DNA and pressure-cooker used.

Sonication: Where available, specialized ultrasonic disintegrators (sonicators) with a probe to put into the DNA (5–50 µg in 100 µl) in solution in a microcentrifuge tube (standing in ice) are an effective and fast method (taking 5–30 s) to fragment DNA. Check DNA length by gel electrophoresis after fragmentation.

Syringing: DNA in solution (5–20 µg in 100 µl) can be fragmented by passing through an extra fine needle (30 gauge) attached to a syringe. The needles may be difficult to source but are cheap (e.g. Hamilton part number 90030). Pull the DNA 50–100 times backwards and forwards through the needle, and check DNA length by gel electrophoresis. Solutions of more than 0.2 µg µl^{-1} will be very difficult to syringe.

Alkali hydrolysis: Add NaOH to a final concentration of 10 M to the DNA, leave at 60°C for 5–10 min, then ethanol precipitate DNA (*Protocol 4.6*). Check length by gel electrophoresis.

DNase digestion: Digest 100 µg DNA in 500 µl nick translation buffer (50 mM Tris-HCl, pH 7.8; 5 mM MgCl$_2$; 0.5 mg ml^{-1} bovine serum albumin (BSA); see *Protocol 4.1*) with DNase I (deoxyribonuclease I; can be a cheap grade containing RNase) for 30 min to several hours. Check length by gel electrophoresis.

3.3 Testing DNA before labeling

Restriction digests to check DNA, linearize plasmids and isolate plasmid inserts

The key technique for testing probes before labeling is size separation of DNA fragments by submarine agarose gel electrophoresis and fluorochrome staining of the DNA, usually with and without restriction enzyme digestion of the DNA. This can be complemented by measurement of DNA concentration by fluorimetry or spectrophotometry (measuring both the A$_{260}$ for DNA concentration and the A$_{260}$/A$_{280}$ ratio for purity); see Sambrook *et al.* (1989) for details.

Restriction enzyme digestion

Restriction enzyme digestions are used:
1. To check DNA amount, size and purity.
2. To linearize plasmids (cut once so the circular plasmid from the bacterial cell becomes linear). This can be used before labeling by nick translation, random primer or PCR amplification; the step can sometimes be omitted,

particularly for PCR amplification (see labeling *Protocols 4.1–4.4*).

3. To isolate plasmid inserts (the cloned DNA of interest) from the vector (e.g. pUC19) so the cloned sequence can be labeled by nick translation or random primer methods without labeling the plasmid vector.

Protocols for restriction enzyme digestion are supplied with most enzymes, and, along with gel electrophoresis methods, are in any molecular biology protocol book or on the Internet site www.jic.bbsrc.ac.uk. Typical amounts of DNA used for digestion are: testing the quality of DNA, 0.5–1 µg; for cutting before labeling, 5 µg; for isolation of insert from vector, 5–30 µg (depending on the amount needed and length of insert).

Agarose gel electrophoresis of DNA

Gel electrophoresis is the key technique for checking the amount and quality of DNA (Hawcroft 1996 and Martin 1996 present details of the methods). An illustration of a gel is shown in *Figure 3.2*. A minigel apparatus with gels *c.* 80 × 60 mm size and six to eight tracks is convenient to check probe DNA. Typically, aim to load 1 µg of DNA in each track, but amounts between 0.01 and 10 µg can be interpreted. Always include DNA size markers ('ladders') with a known amount of DNA in one or more tracks; it is often helpful to use two different ladders. The

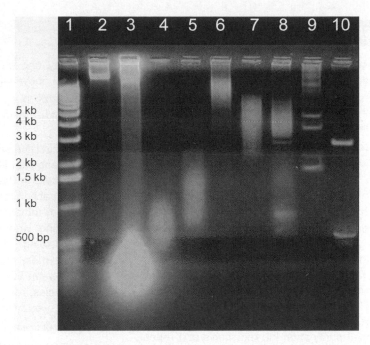

Figure 3.2 Gel electrophoresis. Lane 1: size marker. Lane 2: genomic DNA after isolation and RNase treatment. All DNA is larger than the highest size marker. Lane 3: genomic DNA with RNA contamination (large smear towards bottom of gel). Lane 4: genomic DNA sheared to fragments between 200 and 1500 bp long by autoclaving (1.8 µg was loaded onto the gel in 1 µl of water with 5 µl of loading buffer with dye). Lane 5: genomic DNA sheared to fragments between 1500 and 3000 bp by sonication (1 µg in 10 µl). Lane 6: genomic DNA partially sheared to fragments between 8 and 10 kb long by syringing (1 µg in 5 µl). Lane 7: genomic DNA sheared to fragments of 2–6 kb by syringing. Lane 8: Degraded DNA showing undesirable features such as DNA remaining in the well (top) and weak bands. Lane 9: Isolated pUC18 plasmid DNA showing supercoiled (lower) and open circle (upper) bands. Lane 10: *Bam*HI restriction digest of pUC18 showing the insert of 410 bp (bottom of gel, a 5S rDNA sequence), and an upper band of the plasmid (2.8kb)

DNA in the gel is stained with a fluorochrome (usually ethidium bromide) and the fluorescence examined on a UV transilluminator. Comparison with the markers enables the sizes of DNA fragments to be measured by comparison with distance run, while, with a little experience, comparison of brightness allows concentration to be measured with sufficient accuracy for labeling reactions. In many cases it is advisable to run the DNA both uncut and cut with suitable restriction enzymes (sometimes more than one) for comparison and better interpretation of results. Always photograph the gel for reference in case there are any doubts about the nature of the probe at later stages.

Guidelines are given in *Protocol 3.3* to help interpret results from gel electrophoresis of cut and uncut total genomic, cloned and PCR-amplified DNA stained with ethidium bromide. While much additional analysis is possible, these notes should solve many problems that affect use of DNA as a probe for *in situ* hybridization.

Gel electrophoresis is used:

1. To check quality and quantity of newly isolated/ received DNA (genomic or mini-prep);

2. To check plasmid insert size;

3. To isolate plasmid insert for labeling.

PROTOCOL 3.3 INTERPRETATION OF AGAROSE ELECTROPHORESIS RESULTS

DNA fluoresces bright orange with ethidium bromide staining under UV excitation, seen as white bands on a black background in photographs or computer print-outs (*Figure 3.2*). Interpretation of tracks should consider the following points:

1. No or very little amounts of DNA should remain in the sample loading well. If much remains, it is probably contaminated and will not label well. Repurify the DNA using phenol/chloroform extraction or a commercial purification kit.

2. DNA should run regularly and evenly in the track. If not, the gel may not have been properly dissolved (check the agarose solution is clear before pouring), or the DNA may contain contaminants such as proteins, polysaccharides or a high salt concentration. Purify by precipitation with ethanol (*Protocol 4.6*), phenol/chloroform extraction, or a commercial purification kit, redissolve and repeat electrophoresis.

3. RNA contamination of DNA is seen as a diffuse bright smudge below DNA bands. Treat the DNA with RNase (*Protocol 3.2*).

4. If DNA concentration is much too high, DNA will run unevenly (as it clogs the gel matrix). Reduce concentration and repeat electrophoresis.

5. If no DNA is seen in a track, concentration of DNA is very low; check and consider the following:
   - Check the size markers (ladders) or other tracks are visible to show electrophoresis and staining were correct. Very faint bands may not be visible in a photograph.
   - Check the DNA has not run off the bottom of the gel (run for shorter time, at lower voltage or in higher strength agarose).
   - Possibly precipitate original DNA with ethanol following *Protocol 4.6* and redissolve pellet in a 10 times smaller volume, then repeat electrophoresis.
   - If DNA is still not visible, it may be possible to 'rescue' DNA cloned in a plasmid by recloning, or by PCR amplification (following *Protocol 4.3* without labeled nucleotides), using 5 μl of the solution (or even carry out the PCR reaction in the tube which at one time stored the plasmid).

6. For concentration measurement: Compare sample tracks with molecular weight standards or samples of known concentration. As an approximation, under standard staining conditions, a band containing 0.01 μg of DNA will be at the limits of visibility, one with 0.1–1 μg will be bright, while 5 μg and above will be very bright and often spread out unevenly.

7. GENOMIC DNA shows the following characteristics:
   - Uncut DNA should give a high molecular weight smear, with little DNA visible below 20 kb size makers in the gel. If DNA is below this, it has probably broken following poor storage (e.g. room temperature), poor handling (e.g. Vortex mixing, vigorous pipeting), or contamination.
   - Following digestion with most restriction enzymes, genomic DNA will give a smear of between 500 bp and 20 kb. Depending on the species, restriction satellite bands may be visible overlaying or below the smear.

8. PLASMID DNA shows the following characteristics:
   - Uncut plasmid DNA will normally give two bands corresponding to the super-coiled plasmid (running quickly in the gel) and the open-circle form (nicked and relaxed) which runs more slowly; both molecules are circular so their sizes do not correspond to those of the linear size markers. Additional high molecular weight bands may be present from the bacterial chromosome.
   - The linearized plasmid, cut at one site with an appropriate restriction enzyme, should show a single band corresponding to the size of the plasmid vector plus the cloned DNA insert. When cut with an enzyme that cuts out the insert, a band corresponding to the vector sequence (e.g. 2.7 kb for pUC18/19) and a second band from the insert (typically 150 bp to 5 kb, depending on the size of the DNA cloned) will be visible. Sizes should be measured by reference to the ladder tracks, and the band sizes should correspond to those reported on the plasmid data sheet; length of insert may differ by up to a few tens of base pairs because part of the vector is cut out with the insert. If bands are not visible or are diffuse, the DNA is probably not pure enough, and should be purified by phenol/chloroform extraction or a commercial purification kit. Occasionally, the enzyme may also cut within the insert, when two or more insert bands will be seen. If no insert is seen, the vector may have deleted the inserted DNA fragment.

CHAPTER 4

Labeling the probe

For *in situ* hybridization, nucleotides in the probe DNA or RNA need to be modified (described as labeled, or, in British English, labelled) so that their location can be detected after hybridization with the target DNA. The characteristics of a suitable label include having a chemical structure that is not commonly found in the target preparation, a sensitive detection method that enables detection of only a few label groups, and a means for incorporating the label into the probe DNA or RNA. When *in situ* hybridization was first developed in the late 1960s, probe DNA was labeled with radioactive atoms and signal was detected on photographic emulsion or thin film applied to the slide. Signal takes several weeks to develop and although the method is very sensitive, spatial resolution is poor. Now, non-radioactive labels, where a fluorophore (a chemical group which fluoresces by emission of light of a longer wavelength after excitation with a shorter wavelength of light), a hapten (a chemical group capable of generating antibodies in appropriate conjugates) or other chemical group (the presence of which can be detected) is attached to a nucleotide moiety, are used more widely for labeling DNA and RNA. The non-radioactive labels can be incorporated using standard molecular labeling techniques (*Protocols 4.1–4.5*) and detection uses a color reaction or fluorescence. The simultaneous use of several probes with different labels (or combinations of labels) enables detection of multiple target sequences on a preparation. Improvements in labeling, immunochemistry, microscopy, photographic films, low-light video cameras and image processing have largely overcome early problems with reduced sensitivity compared to radioactive probes. Radioactive *in situ* hybridization protocols are beyond the scope of this book, but are well described by McFadden (1989); they are now used mostly for RNA *in situ* hybridization.

4.1 Labels

For nonradioactive labeling, most researchers use labeled uridine, both in DNA and RNA labels, and leave the remaining nucleotides unlabeled. Use of U is assumed in the labeling protocols, although high GC content probes (such as GC simple sequence repeats) will require labeled G or C to give sufficient signal for detection, and two labeled nucleotides can be used to give higher amounts of label. Although definitive data about the relationship between label incorporation and detection of signal is limited, near-100% replacement of nucleotides with labeled nucleotides in the probe is not an aim of labeling. Steric changes in the probe, affecting both the labeling enzymes and hybridization, probably make high incorporation undesirable. Chains of four to 20 CH_2 groups are usually used to link the label-group and nucleotide to limit these effects (*Figure 4.1*).

Fluorescent detection systems are normally preferred for DNA *in situ* hybridization because the locations of hybridization signals are defined, characteristic and easy to distinguish from dirt or background signal. There have been rapid developments in fluorophore technology since 1990, as the theoretical basis of fluorescence with respect to chemical structure has become better understood and many

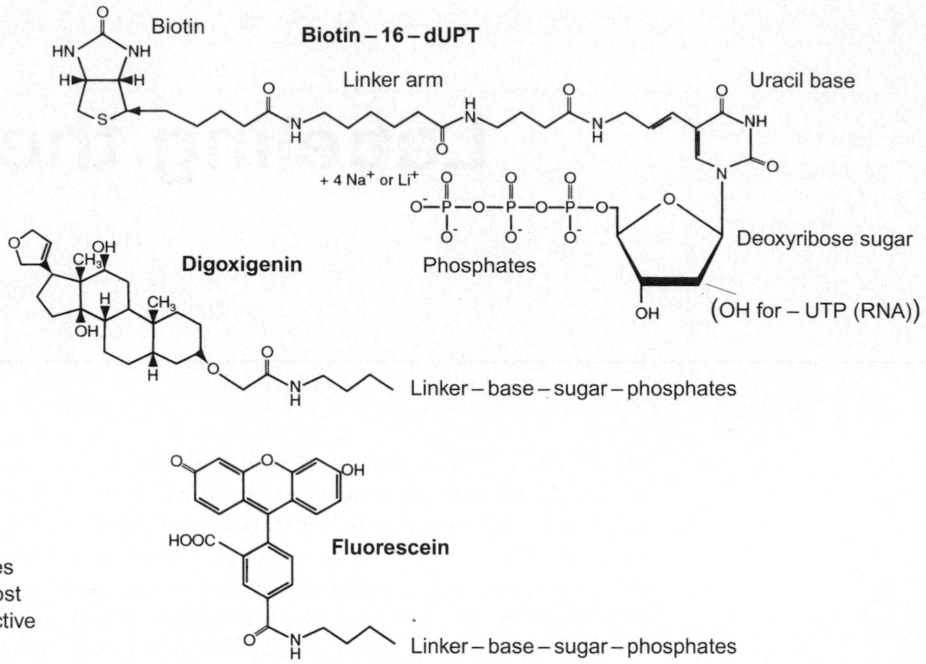

Figure 4.1 Chemical structures of the modified nucleotides most frequently used for nonradioactive labelling of DNA.

fluorophores are now available spanning the visible and invisible spectrum (*Table 4.1*). Several probes can be detected simultaneously using detection reagents with different fluorescent tags. Colorimetric detection produces a stronger, but less spatially defined signal than fluorescence. Hence it is not ideally suited for DNA *in situ* hybridization to chromosomes, but is often used for RNA *in situ* hybridization because the detection of small amounts of RNA at the cellular level is required and because of the greater need for simultaneous through-light imaging of the tissue.

Nonradioactive indirect labels

Many of the first nonradioactive *in situ* hybridization protocols used the naturally occurring molecule biotin (vitamin H) to label the probe and its affinity to avidin for detection of hybridization sites (Langer *et al.* 1981). Now, a variety of labels and anti-hapten antibodies or other detection methods are used. Along with biotin, digoxigenin (a steroid from *Digitalis purpurea*, foxglove) and fluorescein have found by far the widest application as indirect labels for nonradioactive *in situ* hybridization (*Figure 4.1*). Fluorescein, as well as being a direct label (see below) is an excellent hapten with good antigenicity and high affinity for its antibody. These molecules are commercially available linked to nucleotides, and are incorporated into the DNA probe by the labeling techniques presented in Section 4.3. To detect sites of probe hybridization, avidin (or its derivatives, are used for biotin), or anti-hapten antibodies (for digoxigenin, fluorescein and occasionally biotin) linked to fluorescent moieties (see Chapter 9 and *Table 4.1*) or to enzymes which precipitate chromogenic substrates (see Chapter 10).

Direct fluorophore labels

Nucleotides directly conjugated to many fluorophores, emitting most colors under appropriate excitation (*Table 4.1*), are available commercially. With direct fluorophore labels, no detection steps are needed after the post-hybridization washes, so they are quicker and simpler to use than indirect systems, and often show less unspecific background signal than is found with avidin or antibody detection steps. Sensitivity may be more limited than in systems using signal amplification, although anti-fluorophore antibodies can amplify signal (Chapter 9, *Protocol 9.4*) and digital imaging systems (Chapter 11) may overcome problems of low brightness.

Table 4.1 Fluorescent properties of labels and DNA stains used for *in situ* hybridization

	Excitation	Emission
Labels		
Haptens		
Digoxigenin		
Biotin		
Fluorophores[a]		
Cascade blue	400 nm	420 nm
Amino-methyl coumarin[b]	399 nm	445 nm
Cyanine 2	489 nm	506 nm
BODIPY FL	505 nm	515 nm
Alexa 488	490 nm	520 nm
Fluorescein isothiocyanate[b]	495 nm	523 nm
Spectrum green	497 nm	524 nm
Alexa 532	525 nm	550 nm
BODIPY TMR	535 nm	570 nm
Tetramethylrhodamine isothiocyanate[b]	550 nm	570 nm
Cyanine 3	550 nm	570 nm
Alexa 546	555 nm	570 nm
Rhodamine B[b]	560 nm	580 nm
Spectrum orange	559 nm	588 nm
Texas red[b]	595 nm	610 nm
Spectrum red	587 nm	612 nm
Alexa 594	590 nm	615 nm
BODIPY TR	595 nm	625 nm
BODIPY 650/665	650 nm	670 nm[c]
Cyanine 5	649 nm	670 nm[c]
Cyanine 7	743 nm	767 nm[c]
DNA stains		
4′,6-Diamidino-2-phenylindole (DAPI)	358 nm	461 nm
Hoechst 33258 (bis-benzimide)	352 nm	461 nm
Chromomycin A3	430 nm	570 nm
Ethidium bromide	518 nm	615 nm
Propidium iodide	535 nm	617 nm

[a] Most fluorophores are available conjugated to antibodies, avidin and its derivatives, or nucleotides directly. Fluorescent spectra and output can change depending on pH, following conjugation, and manufacturers may use slightly different variants. Further variants of the proprietary color series Cyan (Amersham), Alexa and BODIPY (Molecular Probes) and Spectrum (Vysis) are available.
[b] Often succinimidyl esters are used for conjugation.
[c] Not visible to the human eye and a CCD camera is needed for detection.

Mixtures of labels

Where multiple targets are probed, different labels are attached to each probe and localized simultaneously by the different fluorescence colors (Lichter 1997). (Research is underway to develop reliable, clearly different, noninterfering chromogenic enzyme substrates.) Double-target fluorescence DNA *in situ* hybridization, using red and green fluorescence for detection of probes and a blue-fluorescing counterstain (see Section 4.2) for chromosomes, is used very widely and is little more effort than a single-target hybridization. Double-target methods are almost always worth using when repetitive or chromosome painting probes are being used. Three- or four-target hybridization experiments, with addition of blue or far-red or intermediate fluorochromes (*Table 4.1*), make analysis more difficult but are more informative. (Reprobing slides with additional probes is an alternative option to increase numbers of targets, and particularly useful when targets are present in widely different copy number; Heslop-Harrison *et al.* 1992; see *Protocol 8.4*.)

Ratio-labeled probes, where some probes carry more than one label (mixed either during or after separate labeling reactions), can be used with single-labeled probes to increase the number of targets detected. Where one probe label is detected in red and another in green, further probes with a defined mixture of red and green label (1:1 giving yellow, and other ratios giving yellow-green or orange color) have localized five or more probes simultaneously in both animals (Nederlof *et al.* 1990) and plants (Mukai *et al.* 1993). These results can be analyzed with a conventional fluorescence microscope with suitable filters, although balancing the brightness of the different detection systems and interpreting probes lying next to each other may be difficult. Advanced systems using automated filter wheels or imaging the fluorescence spectra are now available and can detect 24 (all human chromosomes) or more colors simultaneously using five or more labels separately and in different ratios (Liyanage *et al.* 1996; Müller *et al.* 1997; Schröck *et al.* 1996; Speicher *et al.* 1996).

Modified nucleotides

Hapten-*x*-dUTP and Fluorophore-*x*-dUTP

x indicates the length of spacers (normally CH_2 units) between the nucleotide and label moiety. Typically, a spacer of 10–16 units is used where the label is detected by another reaction (because of steric hindrance of attachment of detection reagents), while shorter (down to four units) is used for direct fluorophore labels. Major suppliers include Amersham, Roche, Molecular Probes, Sigma and Enzo.

Most suppliers provide nucleotides in solution at 1 mM; if not, follow the manufacturer's recommendation for dilution or use water to 1 mM final concentration. Very high CG or AT content of the probe might make it necessary to use another labeled nucleotide. dCTP, dGTP and dATP are available.

4.2 Counterstains

Total DNA is usually stained independently of the label (counterstained) to show chromosome morphology and to provide contrast with color used to visualize sites of probe hybridization. DAPI (4',6–diamidino-2-phenylindole), which fluoresces blue under UV excitation (*Table 4.1*), is used most often, being an excellent DNA stain often showing characteristic banding patterns to assist with identification of chromosomes. It also shows negligible excitation and emission at other wavelengths, so does not interfere with excitation and detection of red and green fluorescing hybridization signal. Propidium iodide (PI), showing red fluorescence with green light excitation, is also used as a counterstain, although it is excited and emits over wide wavelength bands and hence interferes with detection of weak hybridization signals. Some probes show sufficient background hybridization that no counterstain is needed or labeled total genomic DNA from the target material can be used as a probe (also providing an internal control for the hybridization; prelabeled DNA can be purchased for this application to human chromosomes from e.g. Vysis). Chromosomes are stained with conventional nonfluorescent stains (e.g. Giemsa) when colorimetric detection is used, but they may quench the fluorescence signals. Depending on the mountant used, chromosomes and nuclei may be seen with phase contrast microscopy.

4.3 Labeling methods

Standard protocols for incorporation of labeled nucleotides into nucleic acids are used for probe labeling. These include nick translation, random primer labeling, PCR-based amplification and label incorporation, end labeling and *in vitro* translation. Below we give protocols we have used successfully to generate probes suitable for *in situ* hybridization. Many companies sell kits for labeling with ready made stock solutions, controls and reliable and easy-to-follow protocols. Companies offer useful advice about their enzymes and kits by telephone or E-mail, in newsletters and through Internet sites. Prelabeled probes are available for a few applications, particularly in human (see Section 4.1).

All labeling protocols start with high quality probe DNA, characterized and tested as in Chapter 3. Scale-up and scale-down of most protocols given here has proven unreliable and therefore we recommend using multiple, small reactions if large quantities of probe are required. Often, kits can be used successfully with smaller amounts of reagent than recommended. Circular DNA (as in plasmids) is not efficiently labeled by random primer labeling or nick translation, and thus should be linearized. Some DNA sequences seem to label poorly by one method, so we use routinely two labeling methods with new probes: hapten- or fluorophore-conjugated nucleotides are not readily incorporated by polymerases and presumably result in incomplete replication, while some DNA sequences have unusual secondary structures which may slow down label incorporation (see Chapter 12).

Nick translation (*Protocol 4.1*) is convenient and generally effective for total genomic DNA or large cloned inserts. We use random primer labeling (*Protocol 4.2*) for small quantities of short (<1 kb) cloned inserts which we cut out of the plasmid vector, run on an agarose gel and purify. The protocol uses less than 200 ng of DNA (and as little as 10 ng) and either double- or single-stranded DNA can be used as a template for the reaction. PCR-based amplification (*Protocol 4.3*) in the presence of labeled nucleotides uses tiny amounts of plasmid or genomic DNA, but yields large amounts of well-labeled probe; however, optimization of the labeling procedure for each probe can be costly in time, nucleotides and enzymes. Oligonucleotides less than 100 bp long are labeled with the enzyme terminal deoxynucleotidyl transferase (TdT) that forms a tail at the 3′ end (*Protocol 4.4*). RNA probes are normally generated by RNA polymerases using promoters and DNA templates cloned into suitable vectors (*Protocol 4.5*). Other labeling methods not based closely on these are available (often from smaller suppliers) or published, including photosensitive labeling reagents, chemical cross-linking methods, unusual DNA polymerases or use of other chemical groups as labels and are only briefly discussed (see end of Section 4.3).

After labeling, the labeled DNA is generally purified to remove the enzyme and unincorporated nucleotides (*Protocol 4.6*) and the incorporation of the label checked by a test blot (*Protocol 4.7*) and, for PCR, by running on an agarose gel. None of the labeling methods is reliable enough, compared to the effort of *in situ* hybridization and detection, that the testing step can be omitted routinely.

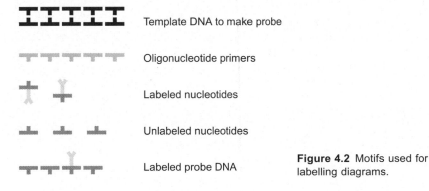

Template DNA to make probe

Oligonucleotide primers

Labeled nucleotides

Unlabeled nucleotides

Labeled probe DNA

Figure 4.2 Motifs used for labelling diagrams.

Nick translation

The nick translation reaction employs two enzymes, deoxyribonuclease I (DNase I) (normally of mammalian origin) and *Escherichia coli* DNA polymerase I (*Figure 4.3*). The DNase I hydrolyzes 'nicks' (1) in each strand of the double-stranded DNA molecule at random (too little nicking can lead to inefficient incorporation of the label and probes that are too long; too much nicking results in probes that are too short). The DNA polymerase I has three activities: an exonuclease function that removes individual base pairs from the nick in the 5′ to 3′ direction, a polymerase function that adds new nucleotides from the 3′ nick copying the template in the opposite DNA strand (2) and a 3′ to 5′ proof-reading activity. Thus, labeled nucleotides in the labeling mixture are incorporated into the newly synthesized DNA (3). Commercially available enzyme mixtures of DNA polymerase I and DNase I are optimized to produce high incorporation of label into DNA in 60–90 min, giving probe lengths of about 200–300 bp.

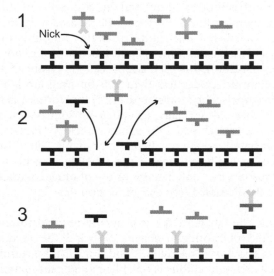

Figure 4.3 The nick translation reaction (see text for details).

PROTOCOL 4.1 NICK TRANSLATION

Follow this protocol, or one supplied with enzyme mixture or labeling kit.

Reagents

$10\times$ nick translation buffer: 500 mM Tris-HCl pH 7.8 (Appendix 1); 50 mM $MgCl_2$; 5 mg ml^{-1} BSA (Note 1).

Unlabeled nucleotide mixture: 0.5 mM dCTP, dGTP and dATP each in 100 mM Tris-HCl, pH 7.5, if dUTP is the labeled nucleotide (Note 2).

Labeled nucleotide mixture: prepare the required mixture in 100 mM Tris-HCl, pH 7.5 (Note 3).
   For digoxigenin: 0.2 mM digoxigenin-11-dUTP (Roche) and 0.4 mM TTP.
   For biotin: 0.4 mM biotin-11 (or 16)-dUTP (e.g. Sigma, Roche) and 0.2 mM TTP.

For direct fluorophore labels: 0.3 mM fluorescein-dUTP or rhodamine-dUTP (e.g. Amersham) and 0.3 mM TTP (see Table 4.1 for alternative labels).

DTT (dithiothreitol), 100 mM solution (77 mg in 5 ml water, store at $-20°C$).

DNA: 1 μg in water or TE (Note 4).

DNA polymerase I/DNase I: this is most conveniently purchased as a mixture at 0.4 units μl^{-1}; and 40 pg μl^{-1} (0.4 milli units μl^{-1}), respectively.

EDTA, 300 mM, pH 8 (Appendix 1).

Method

1. Mix the following in a 1.5 ml microcentrifuge tube (Note 5):

Y μl	Water
3 μl	Unlabeled nucleotide mixture (Note 6)
2 μl	Labeled nucleotide mixture (Note 6)
1 μl	DTT
5 μl	$10\times$ nick translation buffer
X μl	DNA
45 μl =	Total volume

2. Add 5 μl of DNA polymerase I/DNase I solution, mix and centrifuge briefly (Note 7).
3. Incubate at 15°C for 60–90 min (Note 8).
4. Add 5 μl EDTA to stop the reaction and place on ice.
5. Precipitate labeled DNA with ethanol (Protocol 4.6) (Note 9).

Notes

1. Numerous qualities of BSA (50 from Sigma alone) are available; a mid-priced fraction V, essentially globulin-free, specified for use in molecular biology and immunocytochemistry protocols is suitable, such as Sigma B8667, A7638, A3803 or B4287.
2. Do not include the nucleotide that is used as labeled nucleotide. Some protocols use higher concentrations of nucleotides. The mixture may be stored in 100 μl aliquots for 3 months at $-20°C$. Even pure nucleotides degrade over 12 months at $-20°C$ and should be replaced regularly. dNTP sets are available from e.g. Amersham, Roche, Sigma.
3. The ratio of unlabeled to labeled nucleotide might require alteration between 25% and 100% label. Some protocols use higher concentrations of nucleotides.
4. The DNA for labeling must be double-stranded. It can be genomic DNA which needs to be sheared

   to 2–10 kb fragments or linearized and purified plasmid, typically longer than 2 kb.
5. Ideally, add reagents in this order. Biological components should not be added to concentrated buffers.
6. The amount of unlabeled and labeled nucleotides can vary between a final concentration of 0.01 and 0.1 mM.
7. Mix by sharply tapping the tube with the finger several times. Place in a microcentrifuge and spin for 5–20 s to bring the solutions to the bottom of the tube before incubation.
8. Push the tubes through a piece of expanded polystyrene floating in water at 15°C, or use a cooled PCR machine.
9. This step can be omitted for highly repeated or total genomic DNA probes where increased background from unincorporated nucleotides is acceptable.
10. The nick translation protocol is based on Maniatis et al. (1975).

Random primer labeling

Random primer labeling (also known as oligolabeling and random primed labeling) is suitable for smaller amounts of DNA (as low as 40 ng) and shorter fragments than nick translation (see *Figure 4.4*). Single-stranded DNA (1) is amplified with the Klenow fragment of *E. coli* DNA polymerase I ('Klenow enzyme', having a 5' to 3' template-dependent DNA polymerase activity but no nuclease activity) in a (commercially available) random mixture of all possible oligonucleotides (usually hexanucleotides). Since practically all sequence combinations are represented in the hexanucleotide primer mix, the primers bind to the single-stranded template at random positions (2) and provide a primer for semiconservative replication of the DNA strand in the 3' to 5' direction. The new strand is synthesized by the Klenow enzyme in a reaction mixture containing labeled and unlabeled nucleotides (3). The reaction product includes the newly synthesized (labeled) and the template (unlabeled) DNA strands.

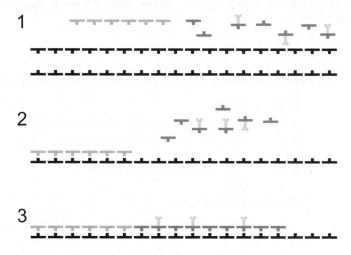

Figure 4.4 Random primer labeling (see text for details).

PROTOCOL 4.2 RANDOM PRIMER LABELING

Follow this protocol, or one supplied with the enzyme or labeling kit.

Reagents

10× Hexanucleotide buffer: 500 mM Tris-HCl, pH 7.2 (Appendix 1); 100 mM MgCl$_2$: 1 mM DTE (dithioerythritol); 2 mg ml^{-1}; BSA with addition of 62.5 A$_{260}$ units ml^{-1}; 'random' hexanucleotide mixture (commercially available).

Unlabeled nucleotide mixture: 0.5 mM dCTP, dGTP and dATP each in 100 mM Tris-HCl, pH 7.5, if dUTP is the labeled nucleotide (Note 1).

Labeled nucleotide mixture: prepare the required mixture in 100 mM Tris-HCl, pH 7.5 (Note 2). For digoxigenin: 0.2 mM digoxigenin-11-dUTP (Roche) and 0.4 mM TTP.

For biotin: 0.4 mM biotin-11 (or -16)-dUTP (e.g. Sigma, Roche) and 0.2 mM TTP.

For direct fluorophore labels: 0.3 mM fluorescein-dUTP or rhodamine-dUTP (e.g. Amersham) and 0.3 mM TTP (see *Table 4.1* for other labels).

DNA: typically 100 ng in water or TE (Note 3).

Klenow fragment of DNA polymerase I: 2 units μl^{-1}, from Amersham, Life Technologies, Sigma D8276 or Roche.

EDTA, 300 mM, pH 8 (Appendix 1).

Method

1. If DNA is double-stranded, place in 10–13.5 μl volume in a 1.5 ml microcentrifuge tube and firmly close lid with tape or a clip. For single-stranded DNA start at Step 3.
2. Bring water in a beaker to boil, switch off heat and float microcentrifuge tube containing DNA in the water. Leave for 5 min and then chill in crushed ice for 5 min (Note 4).
3. Shake the DNA solution to the bottom (Note 5) and mix with the following:

Y μl	Water
2 μl	Unlabeled nucleotide mixture (Note 6)
1.5 μl	Labeled nucleotide mixture (Note 6)
2 μl	10× hexanucleotide buffer (Note 7)
19 μl =	Total volume

4. Add 1 μl of Klenow enzyme solution, mix gently and centrifuge briefly (Note 5).
5. Incubate at 37°C for 6–16 h.
6. Add 2 μl EDTA to stop the reaction.
7. Precipitate labeled DNA with ethanol (*Protocol 4.6*).

Notes

1. Do not include the nucleotide that is used as labeled nucleotide, in most cases dUTP. Some protocols use higher concentrations of nucleotides. The mixture may be stored in 100 μl aliquots for 3 months at −20°C. Even pure nucleotides degrade over 12 months at −20°C and should be replaced regularly. dNTP sets are available from e.g. Amersham, Roche, Sigma.
2. The ratio of unlabeled to labeled nucleotides might require alteration between 25% and 100% label. Some protocols use higher concentrations of nucleotides.
3. Amount of DNA can vary from 40 ng up to 1 μg. Plasmid DNA should be linearized and small

inserts are best cut out from the vector and purified (see Section 3.2). Genomic DNA is also suitable for labeling, but needs to be sheared to 2–10 kb fragments (*Protocol 3.2*).

4. Denaturation must be complete or the primers will be unable to anneal to all parts of the DNA. The quick chilling prevents re-annealing of denatured DNA molecules. Make sure that the tube is tightly closed with a clip or tape to avoid opening when pressure builds up during heating.
5. Mix by sharply tapping the tube with the finger several times. If necessary, place in a microcentrifuge and spin for 5–20 s to bring the solutions to the bottom of the tube.

6. The amount of unlabeled and labeled nucleotides can vary between a final concentration of 0.01 and 0.1 mM.

7. The 10× hexanucleotide buffer is added last to limit re-annealing of the template DNA.

8. This protocol is based on the random primer labeling method by Feinberg and Vogelstein (1983).

Polymerase chain reaction (PCR) labeling

PCR has become a widely used method to amplify and label DNA probes using multiple cycles of DNA denaturation, primer annealing and DNA replication in the presence of a thermostable DNA polymerase, normally *Taq* DNA polymerase. Suitable pairs of synthetic oligonucleotide primers, one complementary to each DNA strand, typically 15–30 nucleotides long, are used to amplify specific defined DNA sequences between the primer pairs. The original DNA and its copies are re-copied multiple times, such that very small amounts of DNA can be used as starting material. DNA sequences, normally up to 2 kb long, inserted into all common cloning vectors including pUC, pBR, pUB or other M13–related vectors can be labeled with the universal M13 sequencing or related primers with the protocol given below. Alternatively, specific DNA sequences can be amplified directly from genomic DNA using suitable primer pairs. The PCR reaction has a number of critical parameters which may be time consuming and expensive to optimize, including the concentrations of template DNA, primers, enzyme, magnesium ions in the buffer and cycling temperatures. Primer design, more advanced techniques and PCR optimization are beyond the scope of this book, but numerous Internet sites (see Additional protocols below and Chapter 3) and books (McPherson and Möller 2000; Newton and Graham 1997) are devoted to the rapidly advancing topic.

In brief, the temperature cycling steps are:

- **Denaturation**. The DNA template is made single-stranded – denatured – by heating to 92–96°C. Lower temperatures risk incomplete denaturation of the target DNA, while higher temperatures destroy the enzyme. Usually, the first denaturation step is longer (1) as it is most critical.
- **Primer annealing**. A pair of specific oligonucleotide primers is annealed to the single-stranded DNA (2) at temperatures between 37 and 70°C, depending on the primer composition, length and homology (see *Protocol 4.3*, Additional protocols and Chapter 13 for calculation programs).
- **DNA synthesis**. The annealed primers allow template-dependent primer extension with a thermostable enzyme (3), usually at between 70 and 74°C, proceeding at about 30 bp s^{-1}.

These three steps are repeated 20–40 times (4).

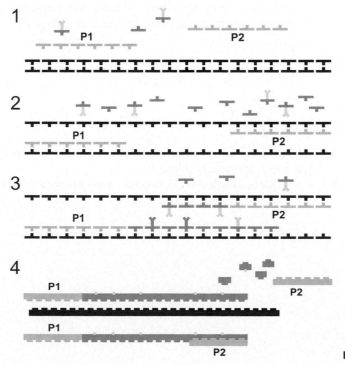

Figure 4.5 PCR labeling.

PROTOCOL 4.3 PCR LABELING OF CLONED DNA SEQUENCES

Reagents

10× PCR buffer: use a 'no Mg^{2+} buffer', usually supplied together with *Taq* DNA polymerase, and containing 100 mM Tris-HCl, pH 8.3 (Appendix 1) and 500 mM KCl and sometimes detergents.

Unlabeled nucleotide mixture: 2 mM dATP, TTP, dCTP and dGTP each in 100 mM Tris-HCl, pH 7.5 (Note 1).

Labeled nucleotide mixture: 1 mM digoxigenin-11-dUTP (Roche), biotin-16 (or -11)-dUTP (Sigma, Roche), fluroescein-dUTP or rhodamine-dUTP (see *Table 4.1* for choice of labels) in 100 mM Tris-HCl, pH 7.5.

Primers: for labeling inserts in M13 and related plasmids with the same multiple cloning site (e.g. pUC18, pUC19 or pBluescript) use the universal M13 forward and reverse sequencing primers: 5'GTA AAA CGA CGG CCA GT 3' and 5'GGA AAC AGC TAT GAC CAT G 3' or variants (Note 2). The stock primer solutions should be diluted to 10 µM.

50 mM MgCl$_2$.

Template DNA: typically 1–10 ng in water or TE (Note 3).

Taq DNA polymerase (1–5 unit µl^{-1}).

Method

1. Mix the following in a microcentrifuge tube suitable for the PCR machine (Note 4):

Y µl	Water
1.5 µl	Unlabeled nucleotide mixture
1.5 µl	Labeled nucleotide mixture (Note 5)
1.5 µl	Primer 1 (Note 2)
1.5 µl	Primer 2 (Note 2)
1.5 µl	MgCl$_2$ (Note 6)
5 µl	10× PCR buffer
X µl	Template DNA
49.5 µl =	Total volume

2. Vortex and centrifuge to bring solution to the bottom of the tube, add 0.5 µl of *Taq* DNA polymerase to each reaction tube. Overlay with mineral oil (e.g. Sigma M3516), two drops per tube (or use a PCR machine with heated lid). Mix again by sharply tapping the tube, and centrifuge to bring to bottom of tube and separate phases.

3. Program your thermal cycler as follows (Note 7):

   93°C 5 min
   94°C 30 s
   55°C 30 s   } 35 cycles
   72°C 90 s
   72°C 5 min
   Hold 4°C

4. Check the PCR product (labeling, size and concentration) by agarose gel electrophoresis (Note 8).

5. Precipitate labeled PCR product with ethanol (*Protocol 4.6*).

Notes

1. Store in 100 μl aliquots for 3 months at −20°C. Some protocols recommend higher concentrations of nucleotides. Even pure nucleotides degrade over 12 months at −20°C and should be replaced regularly. dNTP sets are available from e.g. Amersham, Roche, Sigma.

2. Primers can be synthesized to order or purchased: M13 universal sequencing primers, e.g. Sigma P4165 and P4290 or Strategene #00303 and #300304. T7 and T3 sequencing primers are also suitable for many cloning vectors depending on the insertion site used for cloning. Many companies offer various primers to cater for more unusual insertion sites or other vector systems. Final primer concentration should be between 0.1 and 1.0 mM in the PCR reaction; 0.3 mM is suggested here.

3. Clean uncut plasmid DNA from standard mini-preparations is diluted 1/50 to 1/1000 in water for use in the reaction. Typically, this amounts to 1–10 ng of template DNA (although a wide range, 10 pg to 100 ng, may be used).

4. Many factors affect the results of PCR reactions: amount of template DNA, primer design, enzyme, conditions of amplification and magnesium ion concentration. No amplification or amplification of incorrect template sequences can occur if any is wrong. As the protocol given here describes the amplification of DNA sequences cloned in a plaomid where M13 primers are defined and the annealing at 55°C avoids polymerase pausing at unusual secondary structures of the insert DNA, optimization with multiple reactions and many different controls are not needed. The amount of template DNA is often difficult to estimate and it is advisable to set up test amplification of the desired size of DNA fragment without labeled nucleotide (if necessary adjust the amount of DNA template and other conditions) before repeating with the label in the reaction.

5. Concentration of labeled nucleotide can be important as some haptens may reduce the activity of the enzyme. Possibly, use slightly more labeled nucleotide, especially with direct fluorophores. Biotin, being the smallest molecule of the widely used haptens, is normally accepted best by the enzyme.

6. $MgCl_2$ concentration may need to be adjusted up or down; normally, a final concentration of 1–5 mM is suitable, being 0.5–2.5 mM above that of the total dNTP concentration.

7. The M13 primers perform well at a high annealing temperature (55°C), and normally yield large amounts of a single PCR product. Other primers may need different annealing temperatures and more optimization.

8. PCR amplification and labeling must be tested by gel electrophoresis for amount of amplification, product size and label incorporation (*Figure 4.1*, see Chapter 3 for methods, also see Note 3). Include a marker track with a known amount of DNA and compare the fluorescence brightness by eye after ethidium bromide staining to measure concentration and include a suitable size marker. When amplified by PCR (where probe length is defined by the distance between the primers), labeled probes run more slowly, because of the large labeled nucleotides, than control DNA amplified without label. It is useful to include control amplification without label to see the retardation as a measure of level of incorporation. If a fluorophore direct label has been used, the fluorescence of the labeled DNA fragment, and of the unincorporated nucleotides, will be visible in the appropriate color.

9. Several books discuss aspects of PCR amplification. A helpful internet site for trouble-shooting PCR is available from Promega: www.promega.com/amplification/assistant/.

Additional protocols to label specific sequences directly from genomic DNA

It is possible to amplify and label specific sequences directly from genomic DNA, individual sorted chromosomes or large insert BAC or YAC clones using suitable primers without the need for cloning. For this it is normally advisable to carry out a two-stage amplification: the first without label, then, after checking amplification by electrophoresis, by using 2 μl of the product, with more of the same primers and reaction mixture in a second reaction. Primer design is a complex issue, but helpful computer aids are available, for example at www.genome.wi.mit.edu/cgi-bin/primer/primer3_www.cgi.

Retrotransposon PCR: To amplify and generate labeled retrotransposon probes from any genomic DNA template we have used the primer pairs 5'ACNGCNTTYYTN-CAYGG3'* and 5' ARCATRTCRTCNACRTA3' (making the conserved amino acid sequences, denoted by single letter codes, TAFLHG, and the reverse complement of YVDDML; see Flavell *et al*. 1992). Follow *Protocol 4.3* using 2.5 μl of 50 mM MgCl$_2$ and an annealing temperature of 42°C.

DOP-PCR: The DOP-PCR (degenerate oligonucleotide primed PCR) method (Telenius *et al*. 1992) has been developed to amplify multiple random fragments from genomic DNA, whole chromosomes sorted by flow cytometry or collected by microdissection and large (e.g. YAC) clones (see Chapter 3). DOP-PCR works particularly well with human or mammalian DNA and suitable painting probes have been generated where only very little starting material, for example a few sorted chromosomes or small tumor biopsy is available. We have tried it with plant DNA, but our success was limited perhaps because the degenerate primers do not fit the different codon usage in plants. Set up PCR as described in *Protocol 4.3*, using 0.1–10 ng genomic DNA, 5 μl of 20 μM of the oligonucleotide 5'CCG ACT CGA GNN NNN NAT GTG G 3'* as primer, 2 μl of 50 mM MgCl$_2$. After the initial denaturation of the DNA at 94°C, use five cycles of low stringency amplification (94°C, 1 min; 30°C, 1.5 min; 30–72°C transition, 3 min; 72°C, 3 min), 35 cycles at high stringency (94°C, 1 min; 62°C, 1 min; 72°C, 3 min increasing by 1 s with each cycle) and a final extension of 10 min at 72°C. When testing the PCR products on a gel, a smear of DNA, ranging in size from 200 to 2000 bp should be visible.

Alu- or IRS-PCR: For the specific amplification of human DNA and the generation of human DNA probes from human–rodent somatic hybrid cells or YAC clones (directly from the yeast YAC clone DNA or the YAC DNA isolated after pulse field gel electrophoresis), Alu-PCR or IRS (interspersed repetitive sequence)-PCR protocols were optimized (Lengauer *et al*. 1992). This approach is based on oligonucleotide primers that anneal specifically to human specific sub-sequences within IRSs, for example Alu- or L1–elements and that amplify human DNA sequences between appropriately spaced human Alu- or L1–elements selectively.

Additional protocol to label telomere sequences by PCR

To amplify simple sequence repeats at the telomere, the two strands (TTTAGGG)$_7$ and (CCCTAAA)$_7$ have been synthesized and mixed to allow self-annealing in a PCR amplification (Ijdo *et al*. 1991). High degrees of degeneracy occur because of the lack of proof-reading capacity of the *Taq* DNA polymerase and we prefer using the synthetic or cloned telomere sequence for the detection of true telomeres rather than degenerate dispersed telomere sequences.

---

* Standard abbreviations for nucleotide degeneracies: R = puRine, A or G; Y = pYrimidine, T or C; N = any nucleotide.

End labeling

End labeling (oligonucleotide 3′-end labeling or oligonucleotide 3′-tailing) uses the enzyme terminal deoxynucleotidyl transferase (TdT), a DNA polymerase that will catalyze the addition of nucleotides on to the 3′-OH termini of double- or single-stranded DNA molecules in a template-independent reaction (see *Figure 4.6*). The enzyme prefers to use DNA with protruding 3′-termini as acceptors (1). Blunt or recessed 3′-termini can also be used, although less efficiently, provided that buffers of low ionic strength containing Co^{2+}, Mg^{2+} or Mn^{2+} ions are used. The extent of labeling depends on the number of 3′-OH groups initially present, since these serve as the initiation sites for the nucleotide addition. Large DNA fragments can be cleaved (e.g. with restriction enzymes or DNase I) to increase the number of 3′-OH termini that are available for labeling.

The enzyme will accept modified nucleotides (radioactive nucleotides, hapten- or fluorophore-labeled nucleotides), and if the reaction mixture contains only the modified dNTP, a 'tail' can be produced that is completely labeled (2), although normally a mixture of labeled and unlabeled nucleotides is recommended. End labeling is particularly useful for labeling oligonucleotide probes (i.e. probes less than 100 bp) for which nick translation and random primer labeling methods are not suitable. The speed of the end-labeling reaction is fast, and the kits that are available (e.g. Roche) have optimized conditions to enable the length of the 'tail' to be easily controlled.

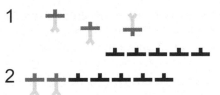

Figure 4.6 End labeling.

PROTOCOL 4.4 END LABELING

Follow this protocol, or one supplied with the enzyme or labeling kit.

Reagents

5× DNA tailing buffer: 1 M potassium cacodylate, pH 7.2 (Note 1); 125 mM Tris-HCl, pH 6.6 (Appendix 1); 1.25 mg ml^{-1} BSA (Note 2).

dATP: 10 mM in 100 mM Tris-HCl, pH 7.5 (Note 3).

Labeled nucleotide mixture: 1 mM digoxigenin-11-dUTP (Roche), or biotin-16 (or -11)-dUTP (e.g.

Sigma, Roche), or fluorescein-dUTP (see *Table 4.1* for choice of label) in 100 mM Tris-HCl pH 7.5.

Cobalt chloride: 25 mM in water.

Template DNA: 250–400 ng in water or TE (Note 4).

TdT enzyme: 30–50 units μl^{-1} (e.g. Amersham, Roche, Sigma).

Method

1. Mix the following in a 1.5 ml microcentrifuge tube:

Y μl	Water
1 μl	dATP
1 μl	Labeled nucleotide
4 μl	Cobalt chloride
4 μl	5× DNA tailing buffer
X μl	Template DNA
19 μl =	Total volume

2. Add 1 μl of TdT, mix gently and centrifuge briefly.
3. Incubate for 15 min at 37°C, or 2–3 h at room temperature.
4. Precipitate labeled DNA with ethanol (*Protocol 4.6*).

Notes

1. Caution: potassium cacodylate is toxic and should be handled with care.
2. Universal restriction enzyme buffers may be used (e.g. Amersham One-phor-All) with substantially increased amounts of enzyme.
3. Any unlabeled or mixture of unlabeled nucleotides can be used. Even pure nucleotides degrade over 12 months at −20°C and should be

replaced regularly. dNTP sets are available from e.g. Amersham, Roche, Sigma.

4. Purified oligonucleotides from 15 to 100 bp long are suitable.
5. End labeling has been described by Tu and Cohen (1980) and modified by Schmitz *et al.* (1991).

In vitro transcription labeling of RNA

In vitro transcription uses bacteriophage RNA polymerase to transcribe RNA from DNA cloned into a suitable RNA transcription vector (usually a phagemid such as pBluescript II from Stratagene) including suitable bacteriophage RNA polymerase promoters adjacent to the cloned sequence (*Figure 4.7*). In the presence of labeled nucleotides, labeled RNA probes (riboprobes) are produced. Most frequently, these are used for hybridization to RNA targets *in situ*, although DNA targets are occasionally used. Different bacteriophage polymerases have different promoters, and the most widely used vectors have different promoters on each side of the cloning site, so single-stranded probes corresponding to each strand of the cloned DNA sequence can be made from one clone. Alternatively, the insert can be cloned in both orientations. In a series of experiments, both strands are labeled and used as probes on different slides; the probe with the same orientation as the transcribed sequence ('antisense') acts as a control and should not show any hybridization, while the 'sense' probe hybridizes to the complementary strand present at mRNA in the tissue.

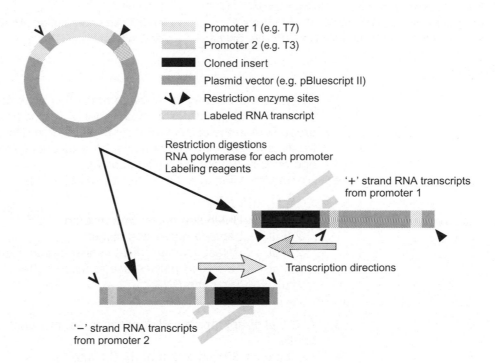

Figure 4.7 *In vitro* transcription labeling.

PROTOCOL 4.5 LABELING RNA BY *IN VITRO* TRANSCRIPTION

Take precaution against RNA degradation. See *Protocol 6.1* (Chapter 6) for preparation of glassware and treatment of all water and solutions used in the protocol. Wear gloves at all times to protect reactions from RNase present on the hands.

Reagents

RNase-free water for all solutions, redissolving pellets and reactions (purchased commercially or DEPC-treated, see *Protocol 6.1*).

$10\times$ RNA polymerase reaction buffer: 400 mM Tris-HCl, pH 8.0 (Appendix 1), 60 mM $MgCl_2$, 100 mM NaCl, 100 mM DTT, 20 μM spermidine

Unlabeled nucleoside mixture: 5 mM CTP, GTP and ATP (ribonucleoside triphosphates) each in 100 mM Tris-HCl, pH 7.5 (Note 1).

Labeled nucleoside mixture: 1 mM digoxigenin-11-UTP (or fluorescein-12-UTP) and 2 mM UTP (Note 2).

RNase inhibitor from human placenta (Roche, Sigma,

Amersham), or cloned (Life Technology), 10 units μl^{-1}.

Template DNA: 0.5–1.0 μg in water or 10 mM Tris-HCl, pH 8 (Note 3).

RNA polymerase (one of T7, T3 or SP6): typically 20 units μl^{-1} (Note 4).

tRNA: 100 mg ml^{-1} in water; Sigma type XXI from *E. coli* Strain W.

DNase I: RNase-free (e.g. Amersham Pharmacia), 7.5 units μl^{-1}.

200 mM carbonate buffer, pH 10.2: 80 mM $NaHCO_3$ and 120 mM Na_2CO_3.

Method

1. Linearize the plasmid DNA with a restriction enzyme that creates a 5′-overhang and generates a linear order of vector, promoter, inserted probe DNA in the correct transcription direction (Note 5 and *Figure 4.7*). Purify template with a commercial kit or by phenol/chloroform extraction and ethanol precipitation (*Protocol 4.6*).

2. Mix the following in a 1.5 ml microcentrifuge tube:

Y μl	Water
2.5 μl	Unlabeled nucleoside mixture
2.5 μl	Labeled nucleotide mixture
2.5 μl	RNase inhibitor (final concentration 1 unit μl^{-1};
2.5 μl	$10\times$ RNA polymerase reaction buffer
X μl	Template DNA
24 μl =	Total volume

3. Add 1 μl of appropriate RNA polymerase, mix gently, and centrifuge briefly.

4. Incubate for 30 min to 2 h at 37°C (Note 6).

5. Optionally, remove template DNA, add 1–2 μl of tRNA, 10 units of RNase-free DNase I and water to a final volume of 100 μl. Incubate for 10 min at 37°C (Note 7).

6. Precipitate labeled RNA with using 100 μl 3.8 M ammonium acetate and 600 μl ethanol (*Protocol 4.6*) and resuspend in 50 μl water.

7. Optionally, hydrolyze probe to shorter fragments by adding 50 μl 200 mM carbonate buffer and incubating at 60°C for 20–90 min depending on original template DNA size (Note 8).

8. Precipitate RNA probe with ethanol (*Protocol 4.6*) and resuspend in 50 μl water.

9. Check label incorporation by a test blot (Note 9).

Notes

1. Do not include the ribonucleoside that is used as label, in most cases UTP. Some protocols use higher concentrations of ribonucleosides. The mixture may be stored in 100 μl aliquots for 3 months at −20°C. Even pure ribonucleosides degrade over 12 months at −20°C and should be replaced regularly. RNase-free NTP sets are available from e.g. Amersham, Roche, Sigma.

2. Digoxigenin and fluorescein are most often used as label, as colorimetric amplification methods are available. Biotin is often not suitable as it occurs naturally in many tissues.

3. The template DNA should consist of a 150–1200 bp insert (optimum 500 bp), without poly A tail, in a high copy vector, e.g. pBluescript II (Stratagene) in front of the T7 or T3 promoter. To generate plus and minus probes (sense and antisense probes) with T7 RNA polymerase, clone the template in both orientations using pBluescript II KS and SK. Alternatively, use one vector to generate both probes using T3 and T7 RNA polymerase. The SP6 promoter is used in alternative vectors. A PCR fragment that has the appropriate promoter ligated to its 5′ ends can also be used as template DNA.

4. Choose the polymerase that will amplify the required fragment depending on the promoter used in cloning.

5. For example, for template DNA cloned into the EcoRV site of pBluescript II KS, cut with HindIII for RNA probes to be generated with T7, and EcoRI for probes to be generated with T3. Use an alternative restriction enzyme in the multi-cloning site if these enzymes are present in the insert.

6. Longer incubation will not yield larger amounts of RNA; scale up reaction instead.

7. As amount of labeled RNA transcript usually far exceeds the DNA template, this step can be omitted.

8. This step generates fragments of 150–200 bp that are recommended for best penetration of probes to the target in tissue sections. If penetration of probes is not a problem, this step can be omitted. A formula can be used to calculate the time needed depending on length of the cloned template DNA insert: $t = (L_i - L_f)/(K \times L_i \times L_f)$, where t = time (min), K = rate constant (= 110 bases min^{-1}), L_i = initial length (kb) and L_f = final length (optimum is 150 bp). In practice 20–30 min incubation is needed for 200–400 bp probes; 60 min for 400–500 bp and 80–90 min for 600–1200 bp probes.

9. Test incorporation of digoxigenin with a test blot (*Protocol 4.7*) or test by removing an aliquot before the DNase treatment Step 5 by agarose gel electrophoresis. An RNA band of the correct size approximately 10-fold more intense than the plasmid band should be seen.

10. Reference: Davenloo et al. 1984.

Chemical labeling

Recent work with *in situ* hybridization has almost always used enzymatic methods to incorporated labeled nucleotides. Methods to add the labeling groups directly to DNA, usually using reactive chemicals, are efficient, can label large amounts of probe, and are often cheap. However, procedures may be difficult, hazardous, use inconveniently large-scale, chemistry-oriented methods, and the label groups attached to the DNA may use nonstandard detection methods. Nevertheless, we foresee that these methods will become more widely used as suitable chemicals and detection methods are developed. It is also likely that new haptens, fluorochromes and perhaps other label groups with increased ease and sensitivity of detection will become available.

- *Photolabeling.* A number of light-sensitive compounds are available that can be used to label DNA and RNA; these include photobiotin and photodigoxigenin.
- *2-Acetylaminofluorene (AAF).* Both double-stranded and single-stranded DNA and RNA can be chemically labeled with AAF, which is highly immunogenic (Landegent *et al.* 1984). AAF is introduced into the nucleic acid by reacting with N-acetoxy-2-acetylaminofluorene.
- *Sulfonation.* Single-stranded DNA probes can be chemically labeled by the insertion of a sulfone group at the C-6 position of cytosine residues. The reaction is catalyzed by sodium bisulfite. The resulting sulfone group is stabilized by the substitution of the amine group on C-4 of cytosine with methoxyamine to produce a sulfonate derivative of cytosine that is highly immunogenic. Approximately 10–15% of the cytosine residues become sulfonated during this reaction. A kit is available from FMC Bioproducts (Chemiprobe kit) for the labeling and subsequent detection of DNA in this way.
- *Mercuration.* Mercury can be incorporated at the C-5 position of pyrimidine bases in nucleic acids (Hopman *et al.* 1987). By varying the incubation time (usually 8–16 h), the degree of mercury modification can be manipulated. Within 8 h, 10–30% of the uracil and cytosine residues are modified in RNA probes and 40–50% of the cytosine bases in DNA.
- *Reactive intermediates.* A number of coupling reagents have been used to cross-link DNA and the label group. Cross-linking fixatives such as glutaraldehyde are used, for example, in the Amersham ECL Southern hybridization method, but do not have high enough labeling efficiency for *in situ* hybridization applications. A promising platinum-containing reactive chemical, linked to biotin and digoxigenin, has been developed by Kreatech, and can label DNA in many forms at 85°C in under 1 h.

4.4 Purification and checking of the probe

DNA precipitation

After labeling of the probe, the labeled nucleic acid is normally precipitated from the labeling solution to remove unicorporated nucleotides, enzymes and salts (which remain in solution and are discarded). The probe is then resuspended in a small volume of water or other solution.

PROTOCOL 4.6 ETHANOL PRECIPITATION

Reagents

3 M sodium acetate (or 4 M lithium chloride), filter
   sterilize and store at room temperature.
96% ethanol at −20°C (Note 1).

70% ethanol at −20°C.
10% acetic acid.

Method

1. For a volume of 50 (or 20) μl from DNA labeling reactions (*Protocols 4.1–4.4*) or 100 μl from RNA probes labeled by *in vitro* translation (*Protocol 4.5*).
2. For DNA probes, add 5 (or 2) μl of 3 M sodium acetate (or 4 M LiCl) and 150 (or 60) μl cold 96% ethanol. For RNA probes, add 5 μl 10% acetic acid to neutralize the carbonate buffer, 10 μl 3 M sodium acetate and 250 μl cold 96% ethanol. It is convenient to transport DNA between laboratories in this state.
3. Precipitate the DNA or RNA in the freezer (−20°C) for 1–2 h or overnight (Note 2).
4. Spin the tubes in a microcentrifuge for 30 min, 12,000*g* (ideally at −10°C, but a centrifuge in a cold room is sufficient) (Note 3).
5. Discard the supernatant and then wash the pellet by carefully adding 0.5 ml of ice-cold 70% ethanol and then centrifuging for 5 min, 12,000*g*.
6. Discard the supernatant by inverting the tube onto an absorbent tissue and leave the pellet until dry (be careful it does not fall out or blow away).
7. Inspect the pellet and check the size and color (Note 3).
8. Dissolve the labeled DNA or RNA in 10–20 μl water (Note 4).
9. Labeled DNA probes are stable and can be stored for long periods at −20°C. RNA probes should be used immediately or can be stored for short periods at −80°C.

Notes

1. 100% ethanol can only be produced by chemical drying, and absorbs water from the atmosphere to give 96% ethanol. 96% ethanol is suitable for the protocol given here, but 100% can also be used.
2. Formerly colder and longer periods were used, but this is no longer considered necessary.
3. It is helpful to have a standard position to place microcentrifuge tubes: with the lid hinge to the upper outside. Then you know to look (and not wipe with tissue etc.) on the hinge site for a pellet.

   A typical pellet from labeled DNA will be visible as a 1×2 mm oval smudge, cream or the color of a fluorophore direct-label.
4. Shake the tube and centrifuge briefly to bring the solution to the bottom of the tube. Leave overnight at room temperature. Do not use a vortex mixer as this will break long DNA molecules. DNA can also be dissolved in TE buffer (Appendix 1; note the EDTA may inhibit subsequent enzyme reactions) or in 100% formamide which may reduce background during hybridization.

Test-blot to check incorporation of label

After labeling, it is necessary to check the incorporation of labeled nucleotides into the probe. For radionucleotides this can be measured using a scintillation counter and in demanding applications (e.g. medical diagnostic probes or quality control of enzymes), such an analysis may be used following simultaneous incorporation of radio-labeled and hapten-conjugated nucleotides. For more routine fluorescent *in situ* hybridization experiments, test incorporation of biotin or digoxigenin by a small test dot-blot following *Protocol 4.7* based on for nonradioactive membrane hybridization with the relevant haptens. Probes can also be size-separated on a gel and transferred to a membrane for Southern hybridization using similar detection methods. Direct-fluorophore labeled probes are checked under a fluorescent microscope by placing 0.5–1 μl of labeled DNA on a glass slide, or size-separating the probe by agarose gel electrophoresis, and exciting it with a transilluminator or the appropriate wavelength with an epifluorescence microscope. A drop including labeled probe will fluoresce in the expected color. If a label test is positive, there is usually no need to measure the concentration of probe further, although it can be measured by fluorimetry, spectrophotometry or comparison with standards following gel electrophoresis (*Protocol 3.3*).

PROTOCOL 4.7   CHECKING BIOTIN AND DIGOXIGENIN LABELED PROBE

Equipment

Plastic forceps for handling membranes (Note 1).

Reagents

Buffer 1: 100 mM Tris-HCl, pH 7.5 (Appendix 1); 15 mM NaCl.

Buffer 2: 0.5% (w/v) blocking reagent (Roche or Amersham) in buffer 1. This is dissolved by heating the solution to 50–70°C for at least 1 h. The prepared solution can then be stored at 4°C for up to 1 month.

Buffer 3: 100 mM Tris-HCl, pH 9.5; 100 mM NaCl; 50 mM $MgCl_2$.

Antibody AP mixture: for the detection of biotin use 1:1000 dilution (final concentration 0.75 units ml^{-1}) of anti-biotin conjugated to alkaline phosphatase (Roche, Sigma, Amersham) in buffer 1. For the detection of digoxigenin use 1:1000 dilution (final concentration 0.75 units ml^{-1}) of anti-digoxigenin conjugated to alkaline phosphatase (Roche) in buffer 1.

Detection reagents (mix the following immediately prior to use), or purchase as pre-mixed and stabilized solution (Life Technologies): 22.5 µl of NBT solution (4-nitroblue tetrazolium chloride, 75 mg ml^{-1} in 70% dimethylformamide). Store frozen in aliquots. 17.5 µl of BCIP (5–bromo-4-chloro-2-indolyl-phosphate, 50 mg ml^{-1} in 70% dimethylformamide). Store frozen in aliquots, 4.96 ml of buffer 3.

Charged nylon membrane as used for Southern hybridization of DNA (e.g. Hybond N^+ from Amersham; off-cut pieces are usually suitable).

Method

1. Cut the membrane to the size required (e.g. 20 × 20 mm).
2. Soak the membrane in buffer 1 for 5 min and then blot dry between filter paper.
3. Micro-pipet small spots of labeled DNA, and controls of previously tested DNA, solution onto the membrane (0.5–1 µl) and leave to adsorb and partly dry for 5–10 min. Write spot identifications on membrane with pencil.
4. Place the membrane in a Petri dish with 4 ml buffer 1 for 1 min, and then place in 4 ml buffer 2 for 30 min. Shake gently.
5. Pour off buffer and distribute 0.5 ml antibody-AP mixture over the membrane (Note 2), cover with a plastic sheet and incubate at 37°C for 30 min, shaking gently.
6. Wash the membrane in buffer 1 for 15 min.
7. Transfer the membrane to buffer 3 for 2 min.
8. Prepare the detection reagents, pour over membrane and leave for 5–10 min in the dark for the color to develop fully.
9. Wash the membrane in water and air-dry (Note 3).

Notes

1. Metal forceps will damage the membrane and cause areas of stain precipitate. Even with plastic forceps, handle gently by corners only.
2. Increase amount if a larger membrane is used.
3. Labeled probe is detected by purple-brown staining of the probe spots. Store test membranes in the dark (e.g. inside a notebook), and mark or describe the signal dots since they will fade over time.

Interpretation of results from probe test dot-blots

Labeled DNA appears as a colored dot on the pale membrane background (*Figure 4.8*). The color may vary between dark brown and dark purple. The intensity of the signal from dots varies, but does not relate clearly to the efficiency of the probe for *in situ* hybridization. Either uniform or no agitation can lead to spread or swirling of signal from the dot (*Figure 4.8(b)*, 6 to 8). There are no differences in signal between the digoxigenin and biotin labeled probes (*Figure 4.8(b)*, 8B is biotin; other probes are digoxigenin). Damage to the membrane from forceps etc. appears as large scratches or areas of stain (at the corners of *Figure 4.8(a)* and near 1 and 2 in *Figure 4.8(b)*); the point of the pipet used to load the sample is also often visible. In *Figure 4.8(a)* spots 1, 2, 3, 7 and 9 are well labeled and suitable for *in situ* hybridization. No label is seen with 5, so this should be discarded. 4, 6 and 8 show weak signal, and could be discarded or tested on unimportant material. In *Figure 4.8(b)*, 6, 7 and 8 are well labeled, while 2 and probably 1 and 3 are unsuitable for *in situ* hybridization.

(a)   (b)

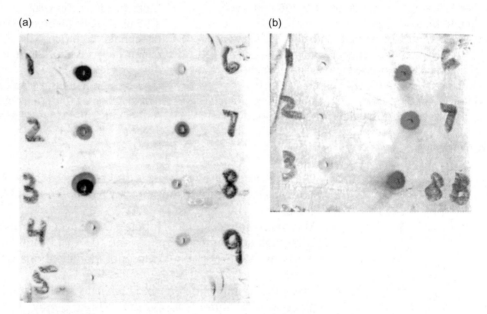

Figure 4.8 Dot blots (twice full size) used to test probe labeling. Brown to dark purple colorimetric precipitates show positions of spots of successfully labeled probe. Numbers are written in pencil on the membrane before treatment.

CHAPTER 5

Preparation of chromosome spreads

An ideal method of tissue preparation ensures both good specimen morphology and that the target molecules are in the optimum state for probe access and hybridization. DNA:DNA *in situ* hybridization is usually carried out on chromosome spread preparations where chromosomes and nucleis are released from cells and spread on a glass microscope slide. This method yields well separated and enlarged chromosomes with good morphology which can be analyzed in through-light or fluorescence microscopes. As precipitating fixatives including alcohol: acetic acid are used, many proteins are destroyed and access of probes to the DNA within the spread chromosomes is very good. However, spreading loses cellular morphology and the three-dimensional organization of the nuclei and chromosomes. Spread preparations are used for mapping and analyzing the long-range organization of DNA sequences, and can be made from mitotic or meiotic material or DNA depending on the resolution required. Methods for spreading chromosomes are widely applicable to plants, animals and fungi, and the following protocols can be adapted to suit all kinds of material.

Methods for fixing, embedding and cutting sections, making whole mounts or cytospin preparations, used for most RNA *in situ* hybridization experiments and three-dimensional DNA:DNA *in situ* hybridization studies are described in Chapter 6.

High quality chromosome preparations are required for the best DNA:DNA *in situ* hybridization. In particular, remains of cytoplasm and other cellular and cell wall material will reduce the *in situ* hybridization signal and generate high levels of background (see also Chapter 12). Ideally, cells and nuclei should be spread to a single layer and there should be little or no contact between cells and nuclei. Individual chromosomes from metaphase cells should be spread apart sufficiently so their number and morphology can be assessed easily. Nevertheless, useful results may come from material that is far from ideal; molecular cytogenetic methods have enabled for the first time cytogenetic analysis of material with no possibility of making metaphase preparations. Interphase nuclei from non-dividing tissues are often used to determine ploidy or aneuploidy with specific DNA probes and several hundred publications describe *in situ* hybridization to human cancers and amniotic fluids. Researchers have been able to analyze 30-year-old biopsy material from wax sections, necrotic tissue cultures and biopsies, or highly differentiated cells using *in situ* hybridization.

Useful protocols about making chromosome preparations from different cells and for different purposes including karyotype analysis and chromosome banding are given by Macgregor and Varley (1988, for many animals), Schwarzacher and Wolf (1974, for human), Verma and Babu (1995, for human) and Fukui and Nakayama (1996, for plants).

5.1 The microscope slide

The first requirement for chromosome preparation is the glass slide. Different batches and suppliers of glass clearly give different qualities of *in situ* hybridization results. With the acid treatment below, all slides seem to work successfully. Omitting this step often leads to selective loss of nuclei and metaphase spreads (see *Protocol 5.8* and *Figure 5.1*) and poor *in situ* hybridization results. The soaking modifies the surface properties of the glass so the spread material is not removed during subsequent treatments. Some slide makes (sometimes batches) may not need soaking.

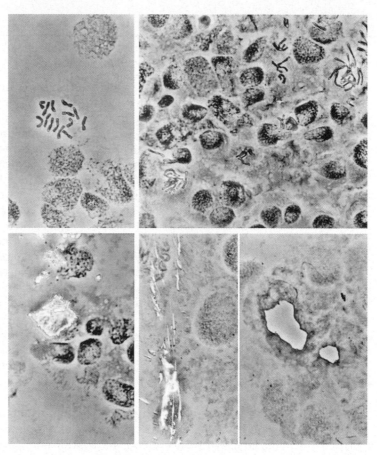

Figure 5.1 Examples of chromosome spreads examined by phase contrast microscopy. These show chromosomes from the wheat species *Triticum monococcum* (2n = 14), but similar preparations can be made from many plants and animals. (**a**) A high quality preparation ideal for *in situ* hybridization: a metaphase and interphase nuclei are apart and little cytoplasm is visible. (**b**) A preparation that has too many cells, with dense cytoplasm between nuclei, distorted chromosomes and dark and contrasty interphase nuclei that are not likely to provide good targets for hybridization. The metaphase index is satisfactory. (**c**) A refractive crystal or particle lying on the slide. There is much cytoplasm around and over nuclei, and, perhaps because of the crystal, neighboring nuclei are poorly spread (dark and contrasty). (**d**) A scratch seen on the slide running through a surface film. The scratch probably arises from careless handling of the slide in solutions or during coverslip removal; heating the slide or more acetic acid treatment might disperse the cytoplasm more. (**e**) A hole in the preparation. If found frequently, the slides may not be pretreated or acid washed.

PROTOCOL 5.1 PREPARATION OF ACID-CLEANED SLIDES

Reagent and equipment

Chromium trioxide solution in 80% (w/v) sulfuric acid (Merck) (often referred to as chromic acid) or 6 M HCl.

1 inch × 3 inch (25 mm × 75 mm) glass microscope slides. One end chemically frosted is convenient for marking with pencil; otherwise legends must be scratched with a diamond pen to survive later stages.

Staining dish with removable slide rack (glass or durable plastic, but not metal; e.g. Sigma S6141).

Method

1. Place slides in a staining dish in acid for at least 3 h at room temperature (Note): safety precautions for strong acids apply.
2. Wash slides in running tap water for 5 min.
3. Rinse slides thoroughly in distilled water.
4. Air-dry at room temperature or 37°C.
5. Place slides into 96% ethanol, leaving for 10 min to 2 days. Remove and dry slides with a lint-free tissue immediately prior to use.

Note

In our laboratory, two racks are left soaking continuously in acid.

5.2 Spreading of plant chromosomes

Any plant tissue containing dividing cells can be used. Usually young root tip meristems (lying immediately behind the root cap) are chosen, but other tissues are possible (in particular developing flower parts, endosperm or leaf and apical meristems). The metaphase index in any meristem can be increased using ice water or spindle-inhibiting drugs (e.g. colchicine or hydroxyquinoline) prior to fixation (*Protocol 5.2*). In some cases, ploidy might be assessed in tissues that do not divide and will rely on interphase analysis. When analyzing meiosis, no metaphase arresting agents should be used as they interfere with the process of meiosis and pairing, and would distort the analysis.

The cell wall of plants, along with the presence of resins and tannins, presents problems that do not occur with animal cells by limiting probe penetration and causing high levels of background. Plant material is usually digested with enzymes to remove the cell wall after fixation. Two methods for preparing mitotic chromosome preparations from plant meristems are described: the squashing method (*Protocol 5.3*) and the dropping method (*Protocol 5.4*). The squashing method usually gives higher quality metaphase spreads and might be the only way to spread large chromosomes from for example wheat, rye or pine, but has the drawback that some chromosomes are distorted. Many *in situ* hybridization experiments using short or low copy target sequences on plant chromosomes have used the dropping method.

Reagents

Metaphase arresting agents:
  Choose one of the following (Note 1) and shake vigorously to aerate before putting in living plant material; except for ice water, the solution should be the same temperature as that where the plants grow to avoid shock.
  (i) *ice water* (for cereals and temperate grasses): put distilled or deionized water in a clean plastic bottle, shake to aerate and keep at −20°C until the water starts to freeze, shake again.
  (ii) *0.05% (w/v) colchicine* (for most plant tissues). Can be stored in the dark at 4°C for several days to weeks.
  (iii) *2 mM 8-hydroxyquinoline* (for dicotyledonous plants, particularly those with small chromosomes such as *Arabidopsis thaliana*). Can be stored in the dark at 4°C for several days to weeks.
  (iv) *α-bromonaphthalene*: store water above liquid α-bromonaphthalene in a bottle (typically 100 ml over 50 ml). Shake, allow to separate, and then pipet out aliquots of the α-bromonaphthalene-saturated water-phase into small vials for treatment of material. Can be stored in a dark bottle indefinitely.
Alcohol:acetic acid fixative: three parts 96% ethanol (or 100% methanol) to one part glacial acetic acid (prepare immediately before use; do not store for more than 30 min as the compounds degrade).

Material

For analysis of metaphase chromosomes, any tissue containing dividing cells can be used: root tips from young seedlings, from newly grown roots at the edge of plant pots or hydroponic culture are all suitable. Alternatively, flower buds, anthers, carpels or leaf or apical meristems can be used (Notes 2 and 3).

For germination of many seeds, put onto filter paper saturated with distilled water at 20–25°C in the dark and leave until roots are about 10–20 mm long. Seeds of trees that grow long single roots are best germinated in a pot of vermiculite, while small seeds are best germinated under sterile conditions on agar minimal medium, e.g. Murashige and Skoog without sugar. Seed suppliers will give advice about germination of difficult species; moving between 4 and 25°C every 3–14 days or smoke treatment often assists germination.

For hydroponic growth, suspend plantlets or bulbs cleaned from soil above an aerated plant nutrient solution (commercial 'complete' plant fertilizers, used at 1/10 the strength recommended as a plant feed are suitable, e.g. Phostrogen); existing roots should be immersed in the solution, but not the plant itself. Root growth is normally initiated within a few days.

Plants established in soil (e.g. trees such as oil palm) may produce actively growing roots within a few weeks when compost is applied on the surface around the stem.

Method

The following steps are carried out in small glass or plastic containers (5–10 ml) or 1.5 ml microcentrifuge tubes. Use generous amounts of solutions: typically 1 ml per specimen. Material is transferred carefully by clean forceps or a pipet (Note 2).

1. To accumulate metaphases, treat excised root tips (5–20 mm long) or other material with one of the metaphase arresting agents as follows (Note 4):

   ice water for 24 h;
   colchicine for 3–6 h at room temperature or 10–24 h at 4°C;
   8-hydroxyquinoline for 1–2 h at room temperature, then 1–2 h at 4°C;
   α-bromonaphthalene saturated water for 2–6 h at room temperature.

2. Quickly blot material and transfer to fixative (Note 5).

3. Fix for at least 16 h at room temperature. If fixed material is to be kept (up to several months), leave for 2 h at room temperature and then transfer to new fixative (or 70% or 96% ethanol) and store at −20°C.

Notes

1. The response to the metaphase accumulation reagents is different from species to species and has to be established by trial and error. Some guidelines for choosing are given; Dolezel *et al.* (1996) discusses alternative reagents, including spindle poisons used as herbicides.

2. It is very important not to expose seedlings, roots and plants during germination and metaphase-arrest to chemicals and fumes, particularly fixatives (e.g. in a cold room also used for chemical storage) and to use clean labware with tight lids (disposable plastic is ideal), clean forceps, and distilled water.

3. Root tips from germinating seeds, and plants grown in controlled conditions, often show waves of cell divisions that may follow internal or environmental rhythms (e.g. light) or correlate with root length. At certain times, there may be no divisions at all, so it may be helpful to make several fixations.

4. Representative times are given. For best results fix material after different times of treatment, experiment with different reagents and check the mitotic index by making chromosome preparations. Treating material for too long in arresting agents, particularly colchicine, results in over-condensation of the metaphase chromosomes which might be desirable for counting chromosomes, but not for *in situ* hybridization where spatial resolution along chromosomes is wanted.

5. Fixative should not be contaminated with water, so careful blotting or an extra rinse in fixative is advised.

Additional protocol for the fixation of meiotic material

Anthers are the normal source of meiotic chromosomes, giving plentiful material. As plant meiocytes progress through meiosis simultaneously within the anthers of one bud, it is advisable to check the stage of meiosis on test anthers with carmine acetic acid or 45% acetic acid before fixation. Bud and anther length can then be used to judge the stage of the other available material. Be aware that some plants undergo meiosis well before buds emerge; in the extreme case of *Hyacinthus* in the autumn prior to the spring they flower. Follow *Protocol 5.2*, omitting the metaphase arresting step (Step 1), and adding 10–50% chloroform to the fixative (Step 2). Fix for shorter times (few hours to overnight) and transfer material to 70% ethanol.

Additional protocol for synchronization of the cell cycle using cold treatment or hydroxyurea

To increase the metaphase index, synchronization of the dividing cell population can be attempted by subjecting seedlings or plants to a period of cold temperatures (generally 24–48 h at 4°C). The plant material is then returned to 20–25°C, and after a recovery of approximately one cell cycle period (dependent on the species; for cereals about 25–30 h), the material is then treated with one of the metaphase arresting agents and fixed as in *Protocol 5.2*. Cold treatments are not suitable for tropical plants.

Hydroxyurea stops the cell cycle reversibly at the beginning of S-phase and has been used to synchronize cell populations in tissues or cultures (see Hadlaczky *et al.* 1983). After germination, transfer seedlings to a new Petri dish with 0.2–2.5 mM hydroxyurea solution saturating filter paper, and incubate in the dark at germination temperature for 15–20 h. Wash seedlings in several changes of distilled water and allow to grow for 4–8 h to recover and start entering metaphase. Follow *Protocol 5.2* for accumulation of metaphases and fixation.

PROTOCOL 5.3 SQUASH PREPARATION OF PLANT CHROMOSOMES

Reagents and equipment

10× enzyme buffer: 40 ml 100 mM citric acid + 60 ml 100 mM tri-sodium-citrate, adjust to pH 4.8. Store stock solution at 4°C. Dilute 1:10 in water for use.

Enzyme solution: 2% (w/v) cellulase from *Aspergillus niger* (Calbiochem, 21947, 4000 units g^{-1}; final concentration: 80 units ml^{-1}) or a mixture of 1.8% Calbiochem and 0.2% 'Onozuka' RS cellulase (5000 units g^{-1}; final concentration: 10 units ml^{-1}) and 3% (v/v) pectinase from *A. niger*

(solution in 40% glycerol, Sigma P4716, 450 units ml^{-1}; final concentration: 13.5 units ml^{-1}). Make up in 1× enzyme buffer. Store in aliquots at −20°C (Notes 1, 2 and 3)

45% and 60% acetic acid: dilute glacial acetic acid with distilled water to the appropriate concentration

Dry ice or liquid nitrogen (Note 4)

Acid-cleaned slides from *Protocol 5.1*

18 × 18 mm (no. 1) coverslips (Note 5)

Dissecting needles and fine forceps

Material

Fixed root tips, buds or other meristem containing material from *Protocol 5.2*.

Method

Each of the following steps, unless stated, is carried out at room temperature in small glass or plastic containers (5–10 ml) or 1.5 ml microcentrifuge tubes. Material is carefully transferred by forceps or, if small roots or buds are used, with a pipet. Alternatively, fluids can be removed with a pipet and replaced with the next required solution.

1. Wash 2–10 root tips or buds twice for 10 min in 2–5 ml enzyme buffer to remove the fixative (until they sink).
2. Transfer material into 1–2 ml enzyme solution and digest at 37°C until the material is soft, usually 45–90 min (Notes 2 and 3).
3. Wash material in 1× enzyme buffer for at least 15 min.
4. Transfer enough material for one preparation (typically one root or small bud) into 45% acetic acid in an embryo dish or small Petri dish for 1–5 min.
5. Make chromosome preparations on an acid-cleaned slide. Under the stereo microscope, in one drop (10–30 µl) of 45% acetic acid (or 60% acetic acid to increase dispersion of cytoplasm), dissect the meristematic tissue by removing as much of the other tissue as possible, e.g. remove the root cap and tease out the cells in the remaining terminal 1–3 mm (Note 6).
6. Apply coverslip to the material without trapping air bubbles. Carefully disperse the material between glass slide and coverslip by tapping the coverslip gently with a needle or flat back of a pencil, and then squash the cells, usually using the thumb with a pressure that just turns the nail white (Note 7).
7. Check the slide under a phase contrast microscope. If not sufficiently flat, squash again.
8. Place the spread slide on to dry ice for 5–10 min (preferred method) or immerse into liquid nitrogen until frozen (about 30 s), then flick off the coverslip with a razor blade (Note 8). Allow the slide to air-dry.
9. Screen slides to choose suitable preparations for *in situ* hybridization (*Protocol 5.8*).
10. Spread preparations can be stored desiccated for up to 3 months at 4 or −20°C.

Notes

1. Most protocols give enzyme concentrations as percentages; we give units and percentages for the products we currently use. Onozuka cellulase is a very pure enzyme and has a constant activity per weight. In general, crude enzyme grades are used: purified enzymes (typically protoplast grades) have different properties and may not give as good preparations for *in situ* hybridization. Batches of the same enzyme product can differ in strength and composition, while distributors may change isolation protocols or suppliers. For example, pectinase from Sigma changes every 2 or 3 years, and they currently sell two (P0690 and P4716) with identical descriptions from different suppliers. With some companies the product number and description may not change although the properties of the product may change greatly.

2. Enzyme mixtures can be reused several times: after use, centrifuge in a microcentrifuge, transfer the supernatant to a new tube, mark for reuse and freeze (not recommended for screening lines of similar material since cells may remain in the solution). Digestion time might need to be slightly increased after each round of use.

3. Sometimes the ratio of cellulase to pectinase needs to be adjusted or different enzymes such as macerozymes, pectolyase or cytohelicase (for meiotic tissue see additional protocol) can be included. The enzyme digestion step needs to be adjusted to the material and species used, by changing the time of digestion; aim at a digestion time of 45–90 min otherwise change the strength of enzyme. Ideally, cell walls should be weakened, so that the cells can be separated easily, and chromosome spreads are clean of cytoplasm. In most cases, the meristematic cells will be digested faster than the non-dividing tissue. The material should remain intact to handle, otherwise the dividing cells are lost into the medium. If material has been fixed for several weeks, the material becomes harder and needs longer digestion.

4. If dry ice or liquid nitrogen are not available, slides can be frozen on a metal plate in a −70°C freezer.

5. Glass coverslips, 18 × 18 mm, of medium thickness are used. As well spread metaphases are often near the periphery of the coverslip, slightly larger coverslips are then used for hybridization and visualization (e.g. 24 × 24 mm) so that all cells can be probed and examined. The coverslips should be free of dirt, but should not be cleaned with alcohol or acid; otherwise material will stick to the coverslip and will be lost when the coverslip is removed (Step 8).

6. It is important to make a monolayer of cells and nuclei and to use the correct amount of fluid that is needed for squashing between the coverslip and glass slide (Step 6). Excess fluid will move all cells to the edge of the coverslip, too little fluid will encourage air bubbles. Thick and too dense material will not spread. Also, be careful to remove any dirt or glass particles (e.g. from writing with a diamond pencil), as they will prevent squashing.

7. Tapping with the needle on top of the coverslip disperses cells, but care needs to be taken not to shear the material by lateral movement of the coverslip. The final thumb pressure to squash the material should be moderately strong, but not abrupt.

8. Alcohol:acetic acid fixative (3:1) at −20°C can be dropped onto or next to the preparation from 10–50 mm immediately after removal of the coverslip, before drying. This seems to disperse cytoplasm, especially in species with small chromosomes.

9. This spreading protocol relies on fixation of the material by precipitation in alcohol:acetic acid followed by chromosome spreading; the method was first developed in plants in the 1930s by C.D. Darlington and L. LaCour (Darlington 1937), replacing complex sectioning methods and allowing very rapid advances in chromosome research. Acetic acid destroys many proteins, making the DNA within the chromosomes accessible to stains or probes. Acetic acid is also used to soften the material prior to spreading and allows chromosomes to stick to the glass slide. Removing of the coverslip follows quick freezing (Conger and Fairchild 1953). The protocol for digestion with enzymes is based on Schwarzacher *et al.* (1980) and produces excellent preparations, even from old fixations.

PROTOCOL 5.4 DROP PREPARATION OF PLANT CHROMOSOMES

Reagents and equipment

10× enzyme buffer: 40 ml 100 mM citric acid + 60 ml 100 mM tri-sodium-citrate adjust to pH 4.8. Store stock solution at 4°C. Dilute 1:10 in water for use.

Enzyme solution: 2% (w/v) cellulase from *Aspergillus niger* (Calbiochem, 21947, 4000 units g^{-1}; final concentration: 80 units ml^{-1}) or a mixture of 1.8% Calbiochem and 0.2% 'Onozuka' RS cellulase (5000 units g^{-1}; final concentration: 10 units ml^{-1}), and 3% (v/v) pectinase from *A. niger* (solution in 40% glycerol, Sigma P4716, 450 units ml^{-1}; final concentration: 13.5 units ml^{-1}). Make up in 1× enzyme buffer. Store in aliquots at −20°C (Note 1).

Alcohol: acetic acid fixative: three parts 96% ethanol (or 100% methanol) to one part glacial acetic acid (prepare immediately before use; do not store for more than 30 min).

Acid-cleaned slides from *Protocol 5.1*.

Material

Fixed root tips, buds or other meristem containing material from *Protocol 5.2*.

Method

1. Wash 5–10 root tips or buds for 10 min in enzyme buffer to remove the fixative. Remove as much of the non-meristematic tissue as possible (i.e. remove root cap and use terminal 3 mm of root tip).
2. Transfer material into enzyme solution in a 1.5 ml microcentrifuge tube and incubate at 37°C until the material is soft and breaks up easily (usually about 2–3 h) (Note 1). The material is gently dispersed with a pipet.
3. Centrifuge for 3 min at 600–800 *g*, remove the supernatant, shake up the pellet (Note 2) and then add 1 ml of 1× enzyme buffer. Mix and leave for 1 min.
4. Repeat Step 3 twice.
5. Centrifuge at 800 *g* for 3 min (Note 3), remove the supernatant, shake up pellet, add 1 ml of fresh fixative and resuspend the pellet with a pipet.
6. Repeat Step 5 twice (Note 4).
7. Centrifuge for 3 min at 800–1000 *g*, discard the supernatant and resuspend the pellet in 50–100 µl of fresh fixative. Drop 10–20 µl of cell suspension onto an acid-cleaned glass slide from 50 mm height, and blow gently (Note 5).
8. Allow the slide to air-dry.
9. The preparations should be screened by phase contrast microscopy (*Protocol 5.8*).
10. Spread preparations can be stored desiccated for up to 3 months at 4 or −20°C.

Notes

1. Adjust the time and the concentration of enzyme to suit the material. Enzyme digestion is normally 1.5–2 times longer than for squash preparations. Alternatively stronger enzyme can be used. See Notes 1, 2 and 5 to squash preparation protocol for advice and use of enzymes.
2. The pellet should be at least 10 µl otherwise it will get lost in subsequent centrifugations and resuspension steps.
3. Centrifugation of the suspension in fixative should be slightly harder, but should not compact the pellet so much that it cannot be resuspended into a smooth suspension. Be careful that you do not shake up the pellet before supernatant is poured off.
4. The suspension can be kept for a few days at 4°C at this point.
5. When the cell suspension is dropped onto the surface of the glass slide, the fixative disperses and evaporates quickly, and forces generated spread the cells, nuclei and chromosomes. This step can be monitored under phase contrast and

adjusted to give well-separated chromosomes in complete metaphase plates. Shaking and blowing the slide increases the spreading forces, while gentle dropping and handling keep the chromosomes together. The density of the cells in suspension is also important for effective spreading. Cells should be abundant, but not overlapping or too close to each other. Repeated centrifugation and resuspension in fixative can help to decrease the amount of dirt in the suspension and allows adjustment of cell density.

6. For the dropping method, a cell suspension, mostly single cells, is required. The method described here follows Ambros *et al.* (1986) and Geber and Schweizer (1988). Fixed material is digested with enzymes and then resuspended in alcohol:acetic acid fixative for dropping. Alternatively, living root tips or suspension cultures are subjected to enzyme treatment to make protoplast suspensions that are then fixed and dropped (Mouras *et al.* 1987; Murata 1983).

Additional protocol for spreading meiotic chromosomes

Plant meiocytes are larger than normal cells and form a strong callose cell wall that is often difficult to remove, particularly after long fixation. We have found that the addition of cytohelicase (2–5%) to the enzyme mixture (Schwarzacher 1997) helps to remove the cell wall and cytoplasm.

5.3 Spreading of insect chromosomes

Polytene chromosomes of *Drosophila* provide excellent targets for single copy *in situ* hybridization because of the 1000-fold duplication of DNA in each polytene chromosome. The method to obtain polytene chromosome spreads is described in detail by Langer-Safer *et al.* (1982) and in Macgregor and Varley (1988). Salivary glands of larvae are excised and fixed in alcohol:acetic acid. Cells are then squashed in 45% acetic acid as described for plant squash preparations (*Protocol 5.3*). The coverslip is removed by freezing and the slides immediately immersed in 95% ethanol. For demonstration purposes, *D. simulans* is easier to use as the larvae have larger glands than *D. melanogaster*.

In grasshoppers and other insects, testis from embryos, larvae or young adults contain both meiotic and mitotic divisions (e.g. Camacho *et al.* 1991; Macgregor and Varley 1988; Vaughan *et al.* 1999). Accumulation of metaphases by colcemid and a hypotonic shock (see mammalian chromosome preparation, *Protocol 5.5*) can be used prior to alcohol:acetic acid fixation and spreading in 45–80% acetic acid under a coverslip or by evaporation of the acid. Aphid chromosome protocols for *in situ* hybridization have been given by Blackman *et al.* (1999).

5.4 Spreading of mammalian chromosomes

In principle, all cell suspension cultures containing mitoses can be used, such as short-term blood cultures and suspensions of adherently growing cultures, for example fibroblast or epithelial cultures. Often the metaphase index is increased by colcemid. An important additional step, a hypotonic treatment, is used with animal cells that is not usually applied to plant material. This step swells the cells and separates the chromosomes prior to fixation.

For prenatal diagnosis or cancer cytogenetics, primary or monolayer cultures are initiated from amniotic fluid or biopsy material. Culturing cells can induce chromosome changes or give selective advantages to certain cells or cell populations. Therefore *in situ* hybridization on uncultured cells is being increasingly employed but has to rely on the analysis of interphase nuclei. Nuclei can be analyzed directly within tissue sections or isolated from frozen material using pepsin and collected using a cytospin (*Protocols 6.3, 6.4* and *6.8*).

PROTOCOL 5.5 MITOTIC CHROMOSOME PREPARATIONS FROM MAMMALIAN SUSPENSION CELL CULTURES

Reagents and equipment

Colcemid: stock solution: 0.01% (w/v) in sterile distilled water.

Hypotonic solution: 0.075 M potassium chloride, or 0.8–1.2% citrate solution (e.g. sodium citrate) or one part culture medium (as used for growing the cells) + three parts distilled water.

Alcohol:acetic acid fixative: three parts 96% ethanol (or 100% methanol) to one part glacial acetic acid (prepare immediately before use; do not store for more than 30 min).

Acid-cleaned slides (*Protocol 5.1*).

Material

Short- or long-term cell culture (Note 1).

Method

Sterile conditions are recommended until cells are fixed (Step 5).

1. Add colcemid stock solution to 10–40 ml culture (1–10 × 10^6 cells) to give final concentration of 0.01%. Incubate for 1–2 h at 37°C or cell growth temperature to accumulate metaphases (Note 2).
2. Transfer cells into one to three 15 ml sterile glass or polypropylene centrifuge tube and centrifuge for 10 min at 1000*g* (Note 3).
3. Carefully pour off the supernatant fluid. Resuspend the cell pellet with the last drop of supernatant by shaking. Do not use a pipet. Slowly add about 10 ml pre-warmed (37°C or room temperature) hypotonic solution and leave to stand for 10–20 min at 37°C or 20–40 min at room temperature (Note 4).
4. Centrifuge at 1000 *g* for 10 min. Pour off supernatant and shake up pellet.
5. Resuspend in about 10 ml of fixative. Add the first 1 ml of fixative dropwise and shake well after each drop. Leave for 10 min at room temperature (Note 5).
6. Repeat Steps 4 and 5 two to five times (Note 6). Cells can be stored in the refrigerator at this point for several days.
7. For spreading, centrifuge the suspension (Step 4) and resuspend in 0.5–1 ml of fixative. Transfer or drop from a height of a few centimeters, one or two drops of well-mixed suspension to a clean slide. Gently shake the slide to distribute cells and blow on them gently. Let dry in an upright position (Notes 7 and 8).
8. View under a phase contrast microscope to check cell density and quality of spread (Note 9). Spread rest of the suspension on slides.
9. Screen slides under a phase contrast microscope (*Protocol 5.8*).
10. Store slides desiccated at 4°C (or −20°C) if necessary.

Notes

1. Subculture long-term cell cultures one day before preparation of chromosomes, so that the maximum growth rate is reached on the day of preparation. For short-term blood cultures, blood is collected in heparin tubes and lymphocytes stimulated with phytohemagglutinin. Diagnostic labs normally separate and count lymphocytes (often from 5ml of blood) to optimize stimulation and growth. Kits are available (e.g. from Difco) using a few drops of blood to yield plenty of slides.

2. Time and concentration needs to be adjusted for different cell types.

3. The pellet should be about 20–50 µl. Too small a pellet can be lost during repeated centrifugation. Some delicate cells will need centrifugation at lower speeds.

4. Hypotonic treatment swells the cells and facilitates untangling of the chromosomes and hence spreading. Time and temperature are critical and vary for cell type and species. Extended treatment and higher temperatures increase swelling of cells. Under-treated cells do not spread well; over-treated cells burst too early and chromosomes are lost.

5. If fixative is added too quickly cells clump and normally cannot be separated again.

6. Centrifugation of the suspension in fixative should be slightly harder than before. Be careful that you do not shake up the pellet before supernatant is poured off. Fixing time can be reduced after the second round, but make sure that overall time in fixative is at least 30 min.

7. When the fixed cell suspension is dropped onto the surface of the glass slide, the fixative disperses and evaporates quickly, and the forces generated spread the cells, nuclei and chromosomes. In some cases, cells may be dropped from 30 cm. Spreading can be monitored under phase contrast and adjusted to give well-separated chromosomes in complete metaphase plates. Shaking and blowing the slide increases the spreading forces, while gentle dropping and handling keeps the chromosomes together. The density of the cell suspension is also important for effective spreading.

8. Some researchers store slides in ice-cold water and drop cells on the wet slides.

9. If chromosomes cannot be separated and spread by increased blowing and shaking, are too dense or sparse, or chromosomes are surrounded by a lot of cytoplasm or other 'dirt', add 10 ml of fixative and repeat Steps 4 and 5.

10. Hypotonic treatment, now used almost universally for swelling cells and disentangling chromosomes was described by Hsu (1952), and was, according to his postscript, an accidental finding. The protocol here is modified from Schwarzacher (1974).

Additional protocol for fibroblast cultures

Fibroblast cultures or other cell types growing on a surface, should be treated with colcemid when still adhering to the surface. A modest trypsin digestion is then used to loosen cells; hit culture bottle sharply to dislodge cells, add culture medium and transfer cells into a 15 ml glass or polypropylene centrifuge tube. Metaphase cells are rounder and will come off the surface more easily, so do not attempt to collect all cells. Centrifuge for 10 min at 350–500*g* and continue with Step 3 of *Protocol 5.5*.

Additional protocol to incorporate bromodeoxyuridine

Bromodeoxyuridine (BrdU) can be added to cell cultures normally synchronized with thymidine before harvesting to label late-replicating DNA. BrdU enhances chromosome elongation and enables the detection of early or late-replicating (R and G) bands using a labeled antibody to BrdU simultaneously with *in situ* mapping (e.g. Lawrence *et al.* 1990).

5.5 Surface spreading of meiocytes

Surface spreading originally involved spreading the contents of cells on top of a hypophase: the surface of water, salt or sugar solution. Cells burst, the chromatin disperses because of the surface tension, and chromatin is picked up with a glass slide or electron microscope grid (Moses 1977). When nuclei at meiotic prophase are spread with this method, the chromatin disperses and the synaptonemal complex (SC), the proteinaceous structure that forms between the entire paired chromosome axis at pachytene becomes visible. SCs can be stained with silver nitrate or electron microscope stains, such as phosphotungstic acid (PTA). Now, most techniques use detergents for bursting the cells and nuclei, and preparations are dried down in the presence of the protein cross-linking fixative paraformaldehyde (Jones and Croft 1986). The chromatin at meiotic prophase is arranged in loops that are attached to the SC and are amenable to *in situ* hybridization allowing the analysis of meiotic prophase chromosome organization (e.g. Albini and Schwarzacher 1992; Moens and Pearlman 1989). *Protocol 5.6* is based on spreading plant meiocytes and includes an enzyme digestion step with cytohelicase that removes the thick callose wall of plant male meiocytes. An additional protocol for animal cells is also given.

PROTOCOL 5.6 SURFACE SPREADING OF PLANT MEIOCYTES

Reagents and equipment

45% acetic acid, orcein or aceto-carmine.

SC enzyme mixture: 0.4–1% cytohelicase from *Helix pomatia* (Sigma or LKB, 3000–4000 units g^{-1}; final concentration: 20–40 units ml^{-1}), 10 mM EDTA (Appendix 1), 6 mM sodium phosphate buffer pH 7.4 (Appendix 1), 0.7% NaCl (Note 1).

Detergent: 0.03–1% (v/v) Lipsol in water (Note 2).

Fixative: in the fume hood, add 4 g of paraformaldehyde (EM grade) and 1.5 g of sucrose to 80 ml water, heat to 60°C for about 10 min, clear the solution with a few drops of 6 M KOH, allow to cool down and adjust the final volume to 100 ml with water. Adjust pH to 8.2 with borate buffer (buy ready-made or make

1.25% boric acid in water, adjust to pH 9 with 1 M NaOH).

Photoflo: 0.1% in water (photographic wetting agent, Kodak; optional).

Acid-cleaned slides (*Protocol 5.1*) or plastic-coated slides (*Protocol 6.2*).

Plastic or glass staining jar (holding slides vertically; e.g. Azlon or Sigma).

Hot plate.

Silver nitrate solution: 0.5 g in 0.5 ml water containing 0.4% paraformaldehyde, filter through a 25 μm filter (Note 3).

Nylon mesh (Nybolt 68GG-243, Schweizerische Gazefabrik, Zürich).

Material

Anthers at meiotic prophase (Note 4).

Method

1. Put SC enzyme mixture, detergent and paraformaldehyde fixative on ice.
2. Determine stage of meiosis by squashing pollen mother cells in 45% acetic acid, orcein or aceto-carmine. Select anthers at pachytene (Note 5).
3. Place one to three anthers on a slide with 35 μl SC enzyme mixture and release pollen mother cells. Incubate for 2 min (Note 6).
4. Add 35 μl of detergent. Mix well and leave for 2 min (Note 7).
5. Add 50 μl of paraformaldehyde fixative, mix well, distribute solution and dry down at room temperature or 40°C on a hot plate in the hood. Leave slides overnight.
6. Wash slides in Photoflo in a staining jar two times for 5 min at room temperature and air-dry (Note 8).
7. Slides can be viewed by phase contrast microscopy or stained with silver nitrate (Note 9).
8. For silver staining, apply 100 μl solution and cover with a nylon mesh. Incubate in a moist chamber at 60°C for 30–60 min. Rinse slides and air-dry. Select slides in the light microscope (Note 10).

Notes

1. Various enzyme mixtures have been published; all use snail enzymes, cytohelicase or glucoronidase, in various buffers that preserve the enzyme activity and ensure that the pollen mother cells are kept in osmotic conditions. One we have also used successfully is 0.4–1% cytohelicase, 1% polyvinylpyrrolidone (PVP, mol. wt 40,000), 1.5% sucrose in water.
2. Other detergents can also be used: e.g. Tween 20, Nonidet, Triton X-100 or certain dish-washing detergents (e.g. Joy), but might require different concentrations.
3. Silver nitrate is a clear solution, but after several

hours stains skin, cloth and benches black and cannot be removed. Be extremely careful when using.

4. Plant meiocytes progress synchronously through meiosis, and trial and error is needed to find them in each species (see step 2 of the protocol). Anther length and color, and also bud size can then be used for future reference.
5. Make sure that the anthers to be used for SC spreading do not come in contact with the acid and are not damaged (some might change color when damaged as a convenient indicator).
6. Time of enzyme treatment needs adjusting to

species used and can vary widely. Meiocytes should be freed into the medium and at least part of the cell walls digested. Possibly also change strength of enzyme solution or use the alternate mixture given in Note 1. Enzyme mixture will only warm slightly during 2 min, but put slide on ice if longer incubations are required.

7. The detergent swells the cells and nuclei. Eventually the nuclei burst and the chromatin is released. This process can be monitored under phase contrast. Meiocytes are round cells, larger than the tapetum or other cells in the anther, with a large nucleus and usually one large nucleolous at the periphery, and if they are spinning (like a *Volvox* sphere), they are ready to burst. The actual release of chromatin often does not happen until the addition of fixative. If nuclei burst too soon, chromatin gets dispersed too much and SCs are destroyed. Concentration of detergent or amount added can be varied, however it is not always easy to distinguish effects of the enzyme and detergent treatment.

8. Photoflo helps to clean slides, but is not necessary. Use distilled water instead.

9. SCs are only visible with phase contrast when very well spread and not in all species. Staining with silver nitrate is necessary and light staining does not interfere much with the *in situ* hybridization. Visualization of SCs, that are often not discernible with DAPI, is then possible with through-light simultaneously to the fluorescent *in situ* hybridization signal.

10. Preparation should turn yellow to light brown. Do not stain too long; SCs should just be visible as light brown lines and phase contrast is needed for screening at low power.

11. The first plant SCs were spread from maize by Gilies (1981), without enzyme digestion. The protocol given here, using enzyme digestion and a detergent for chromatin dispersal, is based on Albini and Jones (1987), Albini and Schwarzacher (1992) and de Jong *et al.* (1989).

Additional protocol for spreading mammalian and insect SCs

Spreading mammalian or insect SCs (see e.g. Jones and Croft 1986; Moens and Pearlman 1989; Speed 1982) is easier than spreading plant SCs, due to the lack of cell walls and because selection of material at the right stage is not needed. Male meiocytes in testis progress asynchronously through meiosis and any biopsy will contain plenty of pachytene material. Sometimes, the most difficult part is obtaining the tissue. Transfer a small piece ($0.5–2$ mm^2) of testis tissue in $50–100$ µl culture medium (e.g. minimum essential medium or RPMI-1640 medium) tease apart; this will release meiocytes from seminiferous tubuli. Transfer $20–40$ µl of medium enriched with meiocytes to a clean slide and continue with step 4 of *Protocol 5.6*.

5.6 Preparation of extended DNA fibers

The extension of DNA fibers to their full molecular length before *in situ* hybridization of one or more labeled probes is a very powerful method to analyze the organization and interspersion of DNA probes at the kilobase level. Impressive electron micrographs of spread DNA fibers were first presented in the early 1970s, and they have been used increasingly for *in situ* hybridization in the 1990s (Brandes *et al.* 1997; Fransz *et al.* 1996; 1998; Haaf and Ward 1994; Heiskanen *et al.* 1994; Heng *et al.* 1992; Houseal *et al.* 1994; Parra and Windle 1993). Techniques to spread DNA out from individual nuclei as extended 'halos' are useful (Gerdes *et al.* 1994), but the extension methods presented here, where fibers are linear, are perhaps easier to analyze and apply to DNA sources from clones and nuclei.

Base pairs of DNA are stacked with a spacing of approximately 0.34 nm in the DNA double helix, so an extended DNA double-helix fiber of 1 kb extends to 0.34 μm (or 2.9 kb μm^{-1}), close to the experimental value of 3.3 kb μm^{-1} obtained from mechanical DNA spreading (*Protocol 5.7C*). Variations are observed between slides because of incorrect removal of extra-nuclear material or faulty mechanical spreading, resulting in inaccurate measurements (Sjöberg *et al.* 1997), so it is best to pool measurements from more than one slide, preferably from more than one experiment.

Figure 5.2 shows how DNA fibers are pulled out from the nuclei.

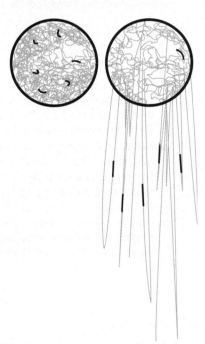

Figure 5.2 Representation of the preparation of DNA fibers where chromatin is dispersed from interphase nuclei (circles), and hybridization target sites (thicker lines) are spread to their full length.

PROTOCOL 5.7 DNA SPREADING – PREPARATION OF EXTENDED CHROMATIN FIBERS

Reagents and equipment

Choose reagents as required:

Nuclear isolation buffer (NIB): 10 mM Tris-HCl, pH 9.5 (Appendix 1), 10 mM EDTA (Appendix 1), 100 mM KCl, 500 mM sucrose, 4 mM spermidine, 1.0 mM spermine and 0.1% (v/v) 2-mercaptoethanol.

For isolating mammalian lymphocytes:
   histopaque or similar density medium for centrifuge separation of lymphocytes.

For isolating plant nuclei:
   liquid nitrogen;
   mortar and pestle;
   10% (v/v) Triton X-100 in NIB;
   nylon mesh with decreasing pore size: e.g. 170, 120, 50 and 20 μm.

STE lysis buffer: 0.5% (w/v) sodium dodecyl sulfate (SDS), 100 mM Tris-HCl, pH 7 (Appendix 1) and 50 mM EDTA (Appendix 1) (Note 1).

Alkali lysis mixture: 50 mM NaOH in 30% ethanol.

Alcohol:acetic acid fixative: three parts 96% ethanol (or 100% methanol) to one part glacial acetic acid (prepare immediately before use; do not store for more than 30 min).

Phosphate-buffered saline (PBS) pH 7.4 (Appendix 1).

TE buffer (Appendix 1).

Acid-cleaned slides (*Protocol 5.1*); poly-L-lysine- or silane-coated slides may also by used (*Protocol 6.2*).

Coverslips, glass, 22 × 40 mm.

Staining jar.

95%, 70% and 50% ethanol.

Proteinase solution: 0.5 M EDTA, 1% SDS with 2 mg ml^{-1} proteinase K.

Low melting point agarose: 3–4% in TE buffer; dissolve by boiling in microwave oven and allow to cool to about 50°C prior to use; for Method C only.

Molds: about 2 mm × 5 mm × 7 mm e.g. blockformers for pulse-field electrophoresis (Biorad); for Method C only.

Phenylmethylsulfonyl fluoride (PMSF): 40 μg ml^{-1}; in TE.

Material

One of the following can be used as source of DNA to give extended fibers:

(a) Nuclear suspension from animal tissue or cell cultures. Protocols for nuclear extraction or high molecular weight DNA isolation required for pulse field electrophoresis are usually suitable. After separation of lymphocyte nuclei, resuspend in NIB at a concentration of 10^5 nuclei ml^{-1} (Note 2).

(b) Nuclei from leaves of young plants. Use the following protocol for isolation (adapted from Liu and Whittier 1994):
   2 g young leaves are ground to a fine powder in liquid nitrogen with a mortar and pestle;

the powder is transferred to a 50 ml centrifuge tube with 20 ml ice-cold NIB. After gently mixing for 5 min the homogenate is filtered through a series of nylon mesh filters with decreasing pore size; add about 1/20 volume of Triton X-100 to the filtrate; centrifuge at 2000g for 10 min at 4°C and resuspend in NIB at a concentration of about 5 × 10^4 nuclei ml^{-1} (Note 2).

(c) Isolated DNA: plasmid, cosmid, BAC, YAC or other clones: isolate high molecular weight DNA and dissolve in PBS or TE buffer.

Method A: using STE lysis

For nuclear suspensions start at Step 1; for isolated DNA, start at Step 3.

1. Centrifuge 50 μl of the nuclear suspension in a microcentrifuge at 3600 rpm for 5 min.
2. Resuspend the pellet gently in 50 μl PBS.
3. Pipet two drops of 1 μl nuclear suspension or isolated DNA solution onto one end of a clean slide and air-dry.
4. Add 10 μl of STE lysis buffer onto the dried down nuclei and incubate for 5 min at room temperature (Note 3).
5. Tilt the slide 45° so that the buffer flows downwards, and let dry down (Note 4).
6. Submerge slides into alcohol:acetic acid fixative in a staining jar for 2 min and air-dry.

7. Incubate slides for 30 min at 60°C (Note 5).
8. It is not necessary to check all slides of a batch, but a selection should be examined by DAPI staining (Note 6).
9. Slides can be stored in a dry box at 4 or −20°C.

Method B: using alkali lysis (for nuclear suspension only)

1. Centrifuge 50 μl of the nuclear suspension in a microcentrifuge at 3600 rpm for 5 min.
2. Resuspend the pellet in fixative.
3. Pipet two droplets of 1 μl cell suspension each onto one end of a clean slide.
4. Before complete evaporation, add 10 μl of 96%, then 70% and 50% ethanol.
5. Add 100 μl of alkali lysis mixture and spread evenly over the nuclei by gently pushing with a coverslip and leave for 40 s.
6. Add carefully 100 μl 96% ethanol, slightly tilt the slide, so that ethanol runs down the slide, and let dry.
7. It is not necessary to check all slides of a batch, but a selection should be examined by DAPI staining (Note 6).
8. Slides can be stored in a dry box at 4 or −20°C.

Method C: using mechanical stretching

For cell suspension start at Step 1; for isolated DNA, start at Step 3.

1. Centrifuge 50 μl of the nuclear suspension in a microcentrifuge at 3600 rpm for 5 min.
2. Resuspend the pellet gently in 50 μl PBS.
3. Place cell suspension or isolated DNA solution in a 1.5 ml microcentrifuge tube and warm to 37°C. Add an equal volume of low melting point agarose at about 50°C (final concentration 1.5–2%). Pipet into small molds with about 10^0 nuclei per block and allow to solidify (Note 7).
4. Remove proteins by incubating blocks in proteinase solution for 24–72 h at 50°C with gentle shaking.
5. Wash blocks four times in PMSF at 50°C for 1 h each.
6. Store blocks in TE at 4°C.
7. Wash block briefly in water and place in a small drop of water towards the end of a microscope slide.
8. Place slide in microwave oven and heat until the agarose is completely but just melted (Note 8).
9. Quickly dip the short side of a 22 × 40 mm coverslip into the molten droplet and pull it once along the surface of the slide, spreading out the agarose uniformly and quickly (Note 9).
10. Submerge slides into fixative in a staining jar for 2 min and air-dry.
11. Incubate slides for 30 min at 60°C (Note 5).
12. It is not necessary to check all slides of a batch, but a selection should be examined by DAPI staining (Note 6).
13. Slides can be stored in a dry box at 4 or −20°C.

Notes

1. Alternatively the buffer can contain only 5 mM EDTA.
2. The nuclear suspension can be mixed with an equal volume of glycerol and stored in $-20°C$ until use.
3. This step disrupts the nuclei; possibly change the time of incubation or concentration of EDTA (Note 1).
4. DNA fibers are stretched when the fluid runs down the slide. The angle of tilting is important. The suspension should run down slowly; if it is very viscous it will need initial higher angle, but will need lowering once it is running. The density of nuclei is critical for good spreading, experiment with different concentrations (see Note 6 for quality control).
5. This sticks DNA firmly to the glass slides so it does not dissolve during later treatments.
6. Add a drop of DAPI (0.05 μg ml^{-1}) in 1× SSC or PBS buffer (Appendix 1) and put on a cover slip before examination with a fluorescence microscope. From a nuclear preparation, remnants of diffuse circular nuclei should be visible, usually with smears of DNA pulling away from them (*Figure 5.1*). Under normal microscopy conditions, bundles of many DNA fibers will be visible. These are not useful for analysis of *in situ* hybridization signal, but will give an idea of the degree of spreading.
7. Make test blocks without cells to test melting properties. Blocks may be stored for a month or more at 4°C in sterile TE buffer.
8. Do not boil. This must be determined by trials with agarose blocks: the position of the slide within the oven, power and time (5–20 s) must be controlled.
9. Use different pressures on the coverslip to get different film thicknesses. The agarose will solidify immediately with the mechanically extended DNA fibers.
10. These protocols are based on Parra and Windle (1993), Heiskanen *et al.* (1994) and particularly Fransz *et al.* (1996; 1998) for plants.

5.7 Quality of chromosome preparation

For *in situ* hybridization, it is advisable to be very rigorous in selecting good slides. For optimum results, the metaphase index should be high, chromosomes need to be well spread, free of cytoplasm and clean (see *Figure 5.1a*). Be careful not to scratch cells, nuclei and chromosomes when handling slides. We routinely discard almost half our preparations from good batches of fixations, and half of all batches of fixations. Different spreads may react differently during the *in situ* hybridization such that a poorly spread metaphase may show no signal while an adjacent good spread will show strong signal. Expertise in chromosome spreading comes with experience, and methods giving good spreads for counting may not be suitable for *in situ* hybridization.

PROTOCOL 5.8 QUALITY CONTROL OF CHROMOSOME PREPARATIONS

5.8

PROTOCOL

When checking your slides consider the following points:

1. Scan slides carefully under phase contrast. Mark the general area on the slide where chromosomes are present underneath using a diamond pencil, otherwise the area may not be visible when placing hybridization solutions onto wet slides. Write down coordinates of good metaphase plates in order to find them again after *in situ* hybridization procedure. Look at the density of spreads, overall number of metaphases, how well spread they are, whether plates are complete (*Figure 5.1b*), and how much cytoplasm is present. Also look for loss of material due to the removal of the coverslip: nuclei may have holes and metaphases may be incomplete and dirt particles and scratches may be present (*Figure 5.1c–e*). Make notes as these help to improve further chromosome preparation experiments and to select the best slides for the *in situ* hybridization experiment, and may determine how to pre-treat slides (*Protocol 8.1*).

2. Possible take photographs: chromosome morphology, and particularly centromeric constrictions, are normally much clearer than after the *in situ* hybridization procedure. However, not all cells will hybridize successfully.

3. Small chromosomes can be stained with DAPI (*Protocol 9.3*) and analyzed under the fluorescence microscope. After viewing, coverslips are removed: (a) with immersion oil on the coverslip, warm slide to 37°C, gently but firmly pull off the coverslip being careful not to get oil on the slide surface and rinse the slide in 70% ethanol; (b) without oil, help the coverslip to fall off in 70% ethanol. Transfer slides through a 70%, 90% and 96% ethanol series (2 min each) and air-dry.

CHAPTER 6

Preparation of sectioned, whole mount or centrifuged material

DNA and RNA *in situ* hybridization to analyze the expression of genes or to detect the presence of bacteria or viruses in infected tissue is most often performed on sections adhered to microscope slides that are, depending on the resolution required, cut from wax or resin embedded material or prepared after freeze drying. Free-floating sections may also be used, and sections can be mounted on electron microscope grids. In some cases, e.g. for the analysis of cell suspensions, cells are centrifuged onto glass slides (using a Cytospin or similar centrifuge apparatus). For three-dimensional analysis of chromosome behavior by DNA:DNA *in situ* hybridization, or RNA detection in whole organism or tissues, whole mounts, thick sections or stacks of thinner sections are analyzed by appropriate methods (electron microscopy, confocal, fluorescent or conventional light microscopy).

As for spread preparation, the fixative and embedding medium must preserve the target sequences and maintain good morphological structure with the additional requirement that three-dimensional relationships of structures must be maintained. Acetic acid, almost universal in spread preparation fixatives, destroys many proteins, making chromosomal DNA accessible to hybridization and helping material to stick to the glass slide, but is devastating for the preservation of the morphology of sections and destroys RNA. For sectioned material, most fixatives are based on the protein cross-linking agents paraformaldehyde and glutaraldehyde in suitable buffers to preserve the target and structure. Too strong or long fixation may reduce accessibility, while dehydration effects may move or coagulate structures within cells. Paraffin wax is the most popular embedding medium as thin sections down to 2 μm can be cut, morphology is well preserved and the wax can be fully removed before hybridization allowing for better penetration of probe. However, if high resolution is required, embedding in resin for the analysis in the electron microscope (EM) is necessary, but is likely to mask some of the target. Greatest sensitivity is often achieved using frozen material (cryopreservation) with minimal chemical fixation, but the morphological preservation is often poor, material is difficult to handle, and certain tissues are not amenable to cryosectioning. When using a Cytospin, the force of the centrifugation helps the cells to stick to the glass surface, thus avoiding the use of acetic acid in fixatives. The choice of the method depends on the material used and the sensitivity of detection required.

Depending on the section thickness and accuracy of sectioning required, sections are cut using a microtome or ultramicrotome (normally moving the embedded or frozen specimen past a static knife and advancing the block forward each cut), a

Vibratome (a rapidly oscillating knife that chops fresh, or sometimes embedded, specimens), or by hand with a razor blade.

The following protocols can be used for preparing sections for DNA and RNA *in situ* hybridization. For RNA *in situ* hybridization, precautions need to be taken to prevent the degradation of the RNA to be detected.

6.1 Precautions against DNA and RNA nuclease activity

For all hybridization experiments, the activities of RNase and DNase enzymes in the specimen itself and solutions used must be controlled. This applies particularly to stages between fixation and detection of hybridization sites, including preparation of the probe (labeling and hybridization protocols). For sample preparation (spreads and sections) with DNA targets, no special precautions are needed as DNA is usually well protected within chromosomes and nuclei. However, RNA is very labile and will degrade very quickly once tissue is damaged or when exposed. Hence, for RNA *in situ* hybridization precautions against RNase activity need to be considered at all times. Always wear disposable gloves and use new, sterile disposable glassware or plasticware where appropriate: it is usually worthwhile to use new reagent and consumable packs (e.g. gloves) specifically for a series of RNA experiments. While DNases are destroyed during the autoclaving of solutions, RNases will survive and diethyl pyrocarbonate (DEPC) is used to treat water or solutions. DNase- and RNase-free water can be purchased for this application.

6.1

PROTOCOL 6.1 TREATMENT OF SOLUTIONS AND GLASSWARE TO DESTROY RNASE ACTIVITY

Purchase RNase- and DNase-free water or follow this protocol. Use new, sterile disposable glassware and plasticware.

Reagents

Diethyl pyrocarbonate (DEPC): e.g. Sigma D5758 (Note 1).
96% ethanol (optional).
3% aqueous hydrogen peroxide (optional).

Method

Treatment of water
1. In the fume hood, add DEPC to pure water to a final concentration of 0.1%.
2. Shake well to dissolve and allow to stand overnight.
3. Autoclave for 15 min at 15 lb in^{-2}, or boil for 30 min in a fume hood.

Treatment of solutions (Note 2)
1. Prepare solutions then, in the fume hood add DEPC to 0.1%. Shake, leave overnight.
2. Autoclave as above.

Treatment of bottles, glass coverslips and other glassware
Bake glassware at 200°C overnight or, if delicate, use the following procedure:

1. Rinse in alcohol.
2. Soak in 3% aqueous hydrogen peroxide for 10 min.
3. Rinse in DEPC-treated water.
4. Dry and protect from dust (typically cover openings with foil).

Notes

1. DEPC is a suspected carcinogen and appropriate care needs to be taken.
2. Tris is destroyed by DEPC and accordingly solutions containing this reagent should be made up in RNase-free glassware using DEPC-treated water and autoclaved.

6.2 Coating of glass slides

Material for *in situ* hybridization and light microscopy is usually mounted on glass slides (plastic surfaces are occasionally used with cultured cells). Chromosome and nuclei spreads do not adhere well to untreated slides and consequently can be lost during *in situ* hybridization. Dirt and grease on the slides can also contribute to background signal. The preparation of acid-washed slides, which are suitable for chromosome spreads, is described in *Protocol 5.1* and may be adequate for sections. To improve adhesion of cells and sectioned material to the glass surface, slides are coated with poly-L-lysine, silane (3-aminopropyltriethoxy-silane, APES) (*Protocol 6.2*) or a proprietary product such as Vectabond (Vector Laboratories). These reagents coat the slide in charged groups which help bind sections to the slide. Some companies (e.g. Sigma and Merck) sell ready-coated slides that work well.

PROTOCOL 6.2 COATING SLIDES WITH POLY-LYSINE OR SILANE

6.2

PROTOCOL

Reagents and equipment

Prepare either of the following solutions on the same day of use:
Silane solution: 1–2% (v/v) 3-aminopropyltriethoxy-silane (APES) in acetone.
Poly-L-lysine solution: 1 mg ml^{-1} poly-L-lysine (MW 300,000) in water.

Acetone.
Acid-cleaned slides from *Protocol 5.1* (Step 5 of alcohol cleaning should be omitted). Alternatively clean slides in 3% laboratory detergent (e.g. Decon) in distilled water overnight, rinse in distilled water and air dry.

Method

1. Soak slides in acetone for 15 min and then bake at 180°C for 2 h.
2. After cooling, either (a) add a small drop (5–10 µl) of poly-L-lysine solution to each slide and, using a baked coverslip (*Protocol 6.1*), draw out into a film to cover the slide (Note 1) or (b) dip slides into silane solution and rinse thoroughly in acetone.
3. Dry slides overnight at 37°C or on a 40°C hotplate (Note 2).
4. Store slides at room temperature or 4°C in a dust-free desiccated environment. May be stored for many months.

Notes

1. If the sections do not adhere well it is possible to aerosol spray the poly-L-lysine to make an even coat on the slide.

2. Possibly, activate APES by placing slides into 2.5% (v/v) glutaraldehyde in 1× PBS (Appendix 1) for 1 h. Wash slides in water and air-dry.

6.3 Paraffin wax sections

Paraffin-embedded material can give excellent preservation of tissue structure and, because the sections form continuous ribbons, serial sections can be collected. Such sections can be used for examination of three-dimensional morphology of whole tissues, or separated onto multiple slides to use with different stains and probes. Paraffin sections, as routinely used in clinical pathology, are suitable for most *in situ* hybridization experiments and there is normally no need for pretreatment of reagents and equipment to remove nucleases as the fixation process will protect tissue nucleic acids from subsequent degradation (see also Chapter 10). However, we recommend fixation using EM grade reagents to give optimum tissue morphology. If low abundance mRNAs are to be detected, or tissue has thick walls or membranes, precautions against RNase are required and care needs to be taken that fixative penetrates quickly. In all cases, fixation should be thorough to preserve morphology and nucleic acids, but not so long that the target becomes unavailable to the probe.

Material is most often fixed in 4% paraformaldehyde and dehydrated prior to embedding in paraffin wax (*Protocol 6.3*). Paraformaldehyde gives good morphological preservation for light microscope (LM) analysis and allows penetration of probes and detection reagents, as it is only a mild cross-linking reagent. Sections (7–10 µm thick, minimum 2 µm) are cut with a microtome and the wax is removed to allow probe penetration (*Protocol 6.4*).

Some material does not embed well in wax, giving poor morphological preservation. In these cases, and when ultra-thin sections (2 µm or less) or high resolution EM analysis are required, embedding in water-soluble media such as acrylic resins (*Protocol 6.5*) or polyethylene glycol (MW 1000 or 1500) are recommended to improve tissue morphology although the tissue sections are harder to handle (Harris *et al.* 1990).

PROTOCOL 6.3 FIXATION AND PARAFFIN EMBEDDING

Reagents and equipment

Paraformaldehyde fixative: in the fume hood, add 4 g of paraformaldehyde (EM grade) to 80 ml water, heat to 60°C for about 10 min, clear the solution with a few drops of 1 M NaOH, let cool down, add 10 ml of 10× PBS (see Appendix 1) and adjust the final volume to 100 ml with water. Adjust to pH 7.2 with 5 M H_2SO_4 (Notes 1 and 2).

Saline solution: 145 mM NaCl (0.85% w/v) in water, put on ice.

100% ethanol (Note 3).

Graded ethanol series in saline solution: prepare 30%, 50%, 75%, 85% and 90% (v/v) ethanol in saline solution. Cool to 4°C.

Erythrocin or safranin: make a very dilute solution in 100% ethanol, giving a watery reddish color (about 0.05%).

Histological clearing agent: Histoclear (Data Diagnostics), Histosol (National Diagnostic), Histolene (Cell Path) or xylene (which is more toxic and flamable).

Paraffin wax: e.g. Paraplast (Shandon), melt at 50–60°C in an oven or water bath (Note 4).

Dissecting needles, forceps and razor blades.

Pasteur pipets (preferably plastic, as fragments from the ends of glass pipets will be embedded with the sample and damage knives).

Material

Plant or animal tissue of interest. Tissue needs to be fixed immediately after dissecting. Bring material before dissection to the fume hood, or transport dissected larger material on ice.

Method

The following steps, unless otherwise stated, are carried out at room temperature in small glass or plastic containers (5–10 ml) or 1.5 ml microcentrifuge tubes using excess fluid volumes (10–20× the volume of material). Fluids are removed with a pipet and replaced with the next required solution. Alternatively material is carefully transferred by forceps or with a pipet.

1. Cut tissue into small pieces, ideally no more than 3 mm in depth. If larger structures need to be fixed, make slits or holes into the surface.
2. Place immediately into fixative in the fume hood (Note 5) for 1–2 h. For vacuum infiltration, place tubes with material (lids open) in a vacuum desiccator and apply and release a vacuum several times. Take care that solution does not boil under reduced pressure. The material should sink once the vacuum is released.
3. Replace with new fixative and leave at 4°C overnight or up to 48 h (Note 6).
4. Wash material in saline solution for 30 min on ice.
5. Dehydrate material in the graded ethanol series for 1–2 h (plant material) or 30 min (animal material) each on ice, and 100% ethanol overnight at 4°C.
6. Replace 100% ethanol with fresh 100% ethanol and leave for 1–2 h (plant material) or 30 min (animal material).
7. Transfer material to erythrocin or safranin solution, leave for 5 min (until red) and wash 3 × 2 min with 100% ethanol (until solution stays clear).
8. In the fume hood, place into 1 part 100% ethanol to 1 part clearing agent for 1–2 h (plant material) or 30 min (animal material) followed by 3 × 1 h in clearing agent.
9. Transfer material into fresh clearing agent and add the same volume of molten paraffin wax. Leave at 40°C overnight.

10. Transfer to molten wax at 60°C and change the molten wax twice a day for 3 days (Notes 4 and 7).
11. Using warmed forceps, place the material into flexible plastic molds containing molten paraffin wax. Arrange material as required. Submerge the molds quickly under water to solidify the wax.
12. Nucleic acids within the wax blocks are stable and blocks can be stored at 4°C for months or years.

Notes

1. We recommend preparation of fresh fixative from EM grade paraformaldehyde powder that has not been in contact with air, immediately before use. However, storage of ready-mixed fixative in the freezer is possible and some protocols dilute the fixative from formalin (a 37% commercially available paraformaldehyde solution), although the impurities will cause poorer morphological preservation. Instead of 1× PBS, 75 mM sodium phosphate buffer (Appendix 1) can be used.

2. Many plant tissues, due to thick cell walls particularly of epidermis cells, show very poor penetration of fixative. Adding of detergents to the fixative will help penetration: 0.1% Tween 20 and 0.1% Triton X-100 is recommended.

3. Use a new bottle of 100% ethanol to ensure that it has not taken up moisture and keep tightly closed. Storing the ethanol with a drying agent is not normally necessary.

4. It is very important that the wax is not contaminated with water.

5. If the material floats, vacuum infiltration is recommended. Sometimes the tissue needs to be weighed down under the surface with a piece of wire gauze, although this is rarely needed if detergents are added to the fixative (Note 2).

6. Fixation at 0–4°C is recommended for RNA work to inhibit endogenous RNases. Room temperature is possible, but shorter times should be used. Time of fixation is critical and needs adjusting to the material. Underfixation may result in poor morphology and loss of nucleic acid and overfixation in masking of target sequences.

7. This can be reduced empirically depending on the tissue. Inadequately embedded material will not cut properly (e.g. the center of the tissue will fall out from the section).

8. This protocol has been adapted from Jackson (1992) and Fobert *et al.* (1994) and includes comments from Tony Warford, Cambridge Antibodies Technologies, Cambridge.

Alternative protocol using automated paraffin wax processing equipment

Standard automated paraffin wax processing equipment can be used for embedding materials suitable for RNA *in situ* hybridization. Fix material as described in *Protocol 6.3*, transfer to 70% alcohol (100% ethanol is not required: industrial methylated spirit or *iso*-propyl alcohol may be used). Automatically process material using an overnight schedule, via ascending alcohols, xylene (or an equivalent clearing agent) into paraffin wax prior to embedding.

PROTOCOL 6.4 CUTTING WAX SECTIONS

Reagents and equipment

Clearing agent: Histoclear (Data Diagnostics),
   Histosol (National Diagnostic), Histolene (Cell Path)
   or xylene (which is more toxic and flammable).
Saline solution: 145 mM NaCl (0.85% w/v) in water.
100% ethanol (Note 1).
Graded ethanol series: prepare 90%, 85%, 70%, 50%
   and 30% (v/v) ethanol in saline solution.

1× PBS (Appendix 1).
Razor blade.
Microtome.
Small paintbrush (as used for water colors).
Coated slides from *Protocol 6.2*.

Material

Paraffin wax block with embedded material from *Protocol 6.3*.

Method

1. Trim the paraffin wax block with a razor blade to give a trapezoid face
   leaving about 2 mm of wax around the tissue. Mount the block with the
   longer of the parallel sides to strike the microtome blade first (Note 2).
2. Cut sections at 6–12 μm thickness (Note 3).
3. Carefully remove ribbons of serial sections with a small paintbrush and
   float them on water on coated slides.
4. Place slides on a hotplate at 40–42°C for a few minutes to allow the
   sections to expand. Drain and leave slides on the hotplate overnight to
   dry.
5. Dewax sections by rinsing slides twice in clearing agent for 10 min each.
6. Rehydrate the sections in the graded ethanol series (100% down to 30%
   ethanol) for 1 min in each followed by incubation in 1× PBS for 5 min
   prior to *in situ* hybridization.

Notes

1. Use a new bottle of 100% ethanol to ensure that it
   has not taken up moisture and keep tightly
   closed. Storing the ethanol with a drying agent is
   not normally necessary.
2. Correct shape of the trimmed block is essential
   for successful cutting and producing ribbons of
   serial sections. Top and bottom face of the trapezoid need to be absolutely parallel and be
   mounted parallel to the knife for sections to form
   a straight ribbon.

3. If sections break up during cutting, material most
   likely has not been embedded properly: either
   dehydration or wax infiltration is probably inadequate. If sections roll up or split along the ribbon,
   the blade is probably dirty or damaged, or the
   cutting angle is wrong. Carefully clean blade with
   96% ethanol, use a new or newly sharpened
   blade and adjust angle.

6.4 Resin sections

Material for examination at high resolution or when serial section reconstructions are required using either LM or EM, is embedded in acrylic resins following fixation in paraformaldehyde or glutaraldehyde (e.g. Binder *et al.* 1986; Leitch *et al.* 1990; McFadden *et al.* 1990; Wachtler *et al.* 1992). While formaldehyde fixation will give poor EM structure, glutaraldehyde will preserve tissue better, but as it cross-links proteins more efficiently than formaldehyde it also will make denaturation of chromosomal DNA and penetration of probes and detection reagents more difficult. Some protocols use a mixture of formaldehyde and glutaraldehyde and in *Protocol 6.6* we describe adding picric acid to the glutaraldehyde fixative. Sections are cut with an ultramicrotome and picked up on pyroxylene-coated gold mesh grids (*Protocol 6.5*). A hydrophobic acrylic resin is used: proprietary types such as LR White (London Resin Company) that is polymerized at 55–65°C are suitable for staining and analyzing material for both EM and LM. If preservation of antigenicity for immunocytochemistry is required, low temperature embedding with Lowicryl K4M (Chemische Werke, Lowi) is recommended (Binder *et al.* 1986; Carlemalm and Villiger 1989).

6.5 | **PROTOCOL 6.5** PREPARATION OF PYROXYLENE-COATED GOLD MESH GRIDS

Reagents and equipment

Butvar solution: 0.15% (w/v) Butvar (Taab Laboratories) in chloroform.
Pyroxylene solution: 4% (w/v) pyroxylene in amyl acetate.
Gold grids: 200 or 400 mesh grids (Note 1).

Method

1. Place gold grids onto filter paper and drop approximately 100 μl of Butvar solution onto each grid.
2. Into a large glass dish containing clean distilled or deionized water place one small drop (approximately 30 μl) of the pyroxylene solution onto the water surface. It should spread into a thin even film.
3. Place the grids onto the film (Butvar surface downwards).
4. Pick up the film by placing filter paper onto the pyroxylene and picking up the filter paper immediately it becomes wet. The grids and film should become attached to the filter paper.
5. Allow the grids to dry (Note 2) and store at room temperature in Petri dish or box.

Notes

1. Gold or nickel grids with square or hexagonal mesh, but not copper grids can be used. A fine mesh is needed as film will break easily after the strenuous *in situ* hybridization procedure.
2. Carbon coat the pyroxylene film if greater stability in the EM is required; this is not usually necessary unless high voltages and low vacuum are used, or the grid is examined and removed from the microscope many times.

PROTOCOL 6.6 — EMBEDDING AND SECTIONING MATERIAL IN ACRYLIC RESIN

Reagents

Fixative buffer: 100 mM phosphate buffer at pH 6.9 (Appendix 1).

Glutaraldehyde fixative: 2% (v/v) glutaraldehyde with 0.2% (v/v) saturated aqueous picric acid is made up freshly in fixative buffer in the fume hood. The fixative can be stored at 4°C for a few days or −20°C for longer. Use a high quality, EM grade of glutaraldehyde, typically supplied as an 8% aqueous solution in sealed glass ampules; lower grades with other aldehydes and impurities will give poorer morphological preservation.

100% ethanol (Note 1).

Graded ethanol series: prepare 10%, 20%, 30%, 50%, 70%, 90% (v/v) ethanol in water.

Acrylic resin: e.g. LR White medium grade (London Resin Company, but available from many suppliers; Note 2).

Gelatin capsules for embedding.

Embedding oven with accurate temperature regulation.

Ultra-microtome with glass or diamond knives: e.g. Reichert Autocut.

Benzyl alcohol solution: 1% (v/v) benzyl alcohol in water.

For LM analysis: coated glass slides from *Protocol 6.2* and xylene-washed copper 3 mm diameter EM grids with single slot (2 × 1 mm).

For EM analysis: coated EM gold mesh grids from *Protocol 6.5* and plastic rings: 2 mm long sections of plastic tube about 3.5 mm diameter to support grids while drying (the tubes used in some wash bottles are very suitable).

Pasteur pipets (preferably plastic, as fragments from the ends of glass pipets will be embedded with sample and damage knives).

Razor blade.

Fine forceps.

Eyelash brush: an eyelash mounted on a small rod (toothpick).

Material

Plant or animal tissue of interest. Tissue needs to be fixed immediately after dissecting. Bring material before dissection to the fume hood, or transport dissected larger material on ice.

Method

The following steps, unless otherwise stated, are carried out at room temperature in small glass or plastic containers (5–10 ml) or 1.5 ml microcentrifuge tubes using 10–20× the volume of fluid to material. Fluids are removed with a Pasteur pipet and replaced with the next required solution. Alternatively, material is carefully transferred by forceps or with a pipet (Note 3).

1. Tissue is cut into small pieces, ideally no more than 1 mm in depth in one dimension.
2. Place immediately into glutaraldehyde fixative in the fume hood (Note 4).
3. Fix for 2 h at room temperature or overnight at 4°C.
4. Wash 2 × 5 min in fixative buffer.
5. Dehydrate through a graded ethanol series, followed by 2 × 100% ethanol, 10 min in each.
6. In the fume hood, preferably on an angled rotating platform, embed in resin by replacing ethanol with resin in the following ratios: 100% ethanol:resin 3:1 (30 min), 1:1 (30 min) 1:3 (30 min) then 100% resin for 2 days changing the solution five times (Note 2).
7. Fill gelatin capsules with resin, transfer and orientate material, and completely fill capsule.
8. Polymerize at 60°C in a small oven in the fume hood for 15 h (Note 5). Blocks can be stored at room temperature for several years.
9. Release embedded material from gelatin capsule by squeezing the capsule or immersing in water to dissolve gelatin.

10. The block containing the material is trimmed to give a trapezoid face and mounted on an ultra-microtome with the long axis to strike the knife edge first (Note 6).

11. Cut to 0.1–0.25 μm thickness. Sections are floated onto a 1% benzyl alcohol solution to re-expand the sections after compression caused by sectioning (Note 7). Use an eyelash brush to manipulate floating sections.

12. For LM analysis, ribbons of consecutive sections are picked up by placing the hole in an EM grid slot over the floating ribbon of sections. The sections are picked up in the drop of liquid and the grid, sections and drop are placed onto a coated glass slide, droplet up. After drying, sections will stick to the glass slides and grids can be shaken off. For EM analysis, use grids with film, pick up sections with the film side of the gold grids, balance droplet up on plastic rings, and allow to air-dry.

Notes

1. Use a new bottle of 100% ethanol to ensure that it has not taken up moisture and keep tightly closed. Storing the ethanol with a drying agent is not normally necessary.

2. Unpolymerized LR White resin is carcinogenic and will penetrate the skin quickly (pathologists use the same protocol to prepare skin biopsies). Work in the fume hood using double gloves; wash hands well with soap and water after the experiment (solvents aid skin penetration). Collect resin and ethanol:resin waste, and polymerize at 60°C before disposal.

3. It is important to avoid contamination with small glass particles or other hard objects that will later damage the microtome knife.

4. We normally do not find penetration of fixative a problem, partly because very small tissue pieces are used. However, adding of detergents (0.1% Tween 20 and 0.1% Triton X-100) to the fixative will help penetration. Also vacuum infiltration can be used. Place tubes with material (lids open) in a vacuum desiccator and apply and release vacuum several times. Take care that solution does not boil under reduced pressure. The material should sink once the vacuum is released. The fixative should then be renewed.

5. Conditions are given for LR White resin. Reduce temperature to 50°C if antigenicity for immunocytochemistry needs to be preserved. Different resins might need different polymerization conditions.

6. Correct shape of the trimmed block is essential for successful cutting and producing ribbons of serial sections. Top and bottom face of the trapezoid need to be absolutely parallel and be mounted parallel to the knife for sections to form a straight ribbon.

7. If a diamond knife is used care must be taken because the benzyl alcohol can dissolve the knife's cement.

8. This protocol follows the method described by Leitch *et al.* (1990).

6.5 Cryosections

Cryosectioning is a fast way to obtain sections; material can be frozen, sectioned and hybridized with a nucleic acid probe on the same day. Material can be fixed either before (Giaid *et al.* 1990; *Protocol 6.7*) or after (Cornish *et al.* 1987; *Protocol 6.8*) freezing and sectioning. The success of each method appears to depend on the material used and we give protocols for each. For mRNA detection, the fixation step usually takes place before freezing to inactivate RNase and to reduce the diffusion of the target sequences.

PROTOCOL 6.7 FIXATION AND PREPARATION OF CRYOSECTIONS OF PLANT MATERIAL

Reagents

Fixative buffer: 100 mM phosphate buffer at pH 6.9 (Appendix 1).

Glutaraldehyde fixative: 2% (v/v) glutaraldehyde (dilute from an EM grade, ready-made 8% aqueous solution) with 0.2% (v/v) saturated aqueous picric acid is made up freshly in fixative buffer in the fume hood. Can be stored at 4°C for a few days (Note 1).

Cryoprotectant and mountant, e.g. OCT compound (Tissue-Tek, Agar Scientific).

Cryostat microtome.

Coated slides from *Protocol 6.2*.

Material

Plant tissue of interest. Tissue needs to be fixed immediately after dissecting. Bring material before dissection to the fume hood, or transport dissected larger material on ice.

Method

The following steps are carried out, unless otherwise described, at room temperature in small glass or plastic containers (5–10 ml) or 1.5 ml microcentrifuge tubes using 10–20× the volume of fluid to material. Fluids are removed with a pipet and replaced with the next required solution. Alternatively material is carefully transferred by forceps or with a pipet.

1. Cut material into small pieces and transfer to glutaraldehyde fixative in the fume hood (Notes 1 and 2).
2. Leave for 2 h at room temperature or overnight at 4°C.
3. Wash 2 × 5 min in fixative buffer.
4. Blot dry material as far as possible.
5. Immerse material in cryoprotectant and freeze at −20°C.
6. Mount frozen tissue block into a cryostat microtome and trim to expose material.
7. Cut thick sections (10–20 μm thick) at −14°C. Ensure the steel blade is sharp and correctly aligned.
8. Pick up sections on a coated slide.
9. Allow sections to thaw and air-dry before using for *in situ* hybridization.

Notes

1. Many plant tissues, due to thick cell walls, particularly of epidermis cells, show poor penetration of fixative. Adding of detergents to the fixative will help penetration: 0.1% Tween 20 and 0.1% Triton X-100 is recommended.
2. If the material floats, vacuum infiltration is recommended. Sometimes the tissue needs to be weighed down under the surface with a piece of wire gauze, although detergents often avoidthat problem. Place tubes with material (lids open) in a vacuum desiccator and apply and release a vacuum several times. Take care that solution does not boil under reduced pressure. The material should sink once the vacuum is released. The fixative should then be renewed.
3. This protocol is taken from Leitch *et al.* (1994a).

PROTOCOL 6.8 PREPARATION OF CRYOSTAT SECTIONS AND FIXATION OF ANIMAL TISSUE

Reagents and equipment

Liquid nitrogen in a plastic or metal Dewar flask.
Liquid nitrogen storage tank.
Embedding medium: e.g. Shandon Cryomatrix™ 67690006.
Isopentane pre-cooled in liquid nitrogen.
TBS 1× (Tris-buffered saline, see Appendix 1).
Paraformaldehyde fixative: in the fume hood, add 4 g of paraformaldehyde (EM grade) to 80 ml water, heat to 60°C for about 10 min, clear the solution

with a few drops of 1 M NaOH, let cool down, add 10 ml of 10× PBS (Appendix 1) and adjust the final volume to 100 ml with water. Adjust to pH 7.2 with 5 M H_2SO_4 (Note 1).
Cork squares: approx. 20 × 20 mm.
Cryotubes or cryogenic vials: e.g. 1.8 ml internal thread; round bottom (Note 2).
Coated slides from *Protocol 6.2*.

Material

Animal tissue of interest. Tissue needs to be processed immediately after dissecting.

Method

1. Dissect representative pieces of tissue, each approximately 5 mm × 5 mm × 2 mm, orientate and place on cork squares covered with embedding medium.
2. Snap freeze for a minimum of 30 s in isopentane pre-cooled in liquid nitrogen then transfer to a cryotube and immerse in liquid nitrogen (Note 3).
3. Mount tissue on a cryostat chuck covered with embedding medium (Note 4).
4. Freeze the assembly by immersing the lower two-thirds of the cryostat chuck into liquid nitrogen.
5. Clamp the chuck into the microtome of the cryostat and trim the tissue until the desired cut surface is exposed.
6. Cut 5 μm sections of the trimmed tissue (Note 5).
7. Mount sections on coated slides and air-dry for 30 min at room temperature (Note 6).
8. Fix slides for 15 min in paraformaldehyde fixative.
9. Wash slides in TBS for 2 × 5 min. Fixed slides can be stored at 4°C for a few days.

Notes

1. We recommend using freshly depolymerized paraformaldehyde, but 10% formol saline (4% formaldehyde, 150 mM NaCl) can also be used.
2. Most tubes are safe for use with vapor phase liquid nitrogen storage. For storage immersed in liquid nitrogen the tubes might need surrounding by Cryoflex.
3. Material in cryotubes can be stored in vapor phase liquid nitrogen.
4. Unless reorientation is required avoid thawing of tissues.
5. When required number of sections are cut, remove the tissue/chuck assembly from the

cryostat. The remaining material can be stored in liquid nitrogen. To protect the cut surface of the tissue from freeze-drying during storage cover with embedding medium and freeze as in Step 2. Detach the tissue from the chuck by cutting the embedding medium between the cork and chuck, return the tissue to its cryotube and store in vapor phase liquid nitrogen.
6. Slides can be stored covered at −70°C. When slides are required, remove from −70°C and thaw in sealed container or dessicator at room temperature for 30 min and fix (Step 8).
7. Protocol supplied by Tony Warford, Cambridge Antibody Technology, Cambridge.

6.6 Whole mount preparations

For the study of whole organisms (e.g. *Drosophila* or vertebrate embryos; Tautz and Pfeiffle 1989; Wilkinson 1992), organs (e.g. young pea embryos; Harris *et al.* 1990), or undisrupted cells (e.g. Visser *et al.* 1998), material up to 1–2 mm in diameter can be fixed using precipitating or cross-linking fixatives. RNA or DNA *in situ* hybridization protocol is then conducted on the whole material or on fresh thick sections. This technique retains the topography of the tissue and allows analysis by confocal microscopy or deblurred optical sectioning (see Chapter 11). Whole mount preparations are used increasingly for the detection of mRNA (see Rosen and Beddington, 1993; Wilkinson 1992), and are also used for the study of the three-dimensional organization of the nucleus in animals and plants (Carter *et al.* 1993, Dietzel *et al.* 1998, Kurz *et al.* 1996, Neves *et al.* 1995b; Rawlins *et al.* 1991).

Problems encountered in whole mount preparations include non-uniformity of fixation, pretreatments, denaturation, and probe or detection reagent penetration. Cells are often permeabilized using detergents or cycles of freezing and thawing, and proteases can be used before or after fixation. Plant cell walls are often partially digested (e.g. with cellulase and pectinase, as described in *Protocols 5.3* and *5.4*). Many animal tissues will need special treatments or bleaching to remove membranes or pigment granules. If antibodies are used to detect the sites of probe hybridization (e.g. anti-digoxigenin), background is reduced by pre-adsorbing the antibodies on control material (not used for the experiment) to remove any non-specific antibody activity. The protocol below is based on whole mount human fibroblasts or small plant structures as used for DNA:DNA *in situ* hybridization, but can easily be adapted for other material and RNA *in situ* hybridization.

PROTOCOL 6.9 WHOLE MOUNT PREPARATIONS FOR DNA:DNA *IN SITU* HYBRIDIZATION

Reagents and equipment

Choose reagents as required:

1× PBS (see Appendix 1).

1× TBS (see Appendix 1).

Paraformaldehyde fixative: in the fume hood, add 4 g of paraformaldehyde (EM grade) to 80 ml water, heat to 60°C for about 10 min, clear the solution with a few drops of 1 M NaOH, let cool down, add 10 ml of 10× PBS and adjust the final volume to 100 ml with water. Adjust to pH 7.2 with 5 M H_2SO_4 (Notes 1 and 2).

Triton X-100 solution: 0.5% (v/v) in 1× PBS.

20% glycerol in 1× PBS.

Liquid nitrogen.

10× enzyme buffer: 40 mM citric acid, 60 mM tri-sodium-citrate pH 4.8. Store stock solution at 4°C. Dilute 1:10 in water for use.

Enzyme solution: 2% (w/v) cellulase from *Aspergillus niger* (Calbiochem, 21947, 4000 units g^{-1}; final concentration: 80 units ml^{-1}) and 3% (v/v) pectinase from *Aspergillus niger* (solution in 40% glycerol, Sigma P4716, 450 units ml^{-1}; final concentration: 13.5 units ml^{-1}). Make up in 1× enzyme buffer (Note 3).

2× SSC: dilute from 20× SSC (Appendix 1).

0.1M HCl.

Glue: cyanoacrylate 'Super-glue' (e.g. Bostik Super Glue 4).

Multiwell slides: glass slides with clear circles (typically 8 mm diameter) surrounded by a hydrophilic coating (e.g. Shandon). Treat slides with APES following *Protocol 6.2*.

Material

Cultures growing on solid support: grow cells on sterile coverslips (26 mm × 76 mm, 170 µm thickness), wash a few times in 1× PBS before starting method. Process coverslips with adhering cells.

Plant tissue: dissect material e.g. root tips or anthers; if larger structures need to be analyzed make cuts or holes into the surface.

Method

The following steps, unless otherwise stated, are carried out at room temperature in small glass or plastic containers (5–10 ml) or 1.5 ml microcentrifuge tubes using 10–20× the volume of fluid to material. Fluids are removed with a pipet and replaced with the next required solution. Alternatively material is carefully transferred by forceps or with a pipet.

Coverslips with adhering cells are treated in small staining jars. Be careful not to let cells dry out during and between steps.

1. Place cover slips with cells or dissected tissue in paraformaldehyde fixative in the fume hood. Plant tissue might need vacuum infiltration: place tubes with material (lids open) in a vacuum desiccator and apply and release a vacuum several times. Take care that solution does not boil under reduced pressure. The material should sink once the vacuum is released. The fixative should then be renewed.
2. Leave for 10–20 min at room temperature or overnight at 4°C (Note 4).
3. Wash material 3 × 5 min in 1× PBS or TBS.
4. For solid tissue, cut sections with a Vibratome or by hand with a razor blade. Stick material to a glass rod using a small drop of glue, cut sections of 40–50 µm thickness and collect under water. Transfer sections to clear areas of multiwell slides and let dry down.

5. Permeabilization of material:
Either (a) cells on cover slips:

(i)   Permeabilize in Triton X-100 solution for 10–20 min.
(ii) Incubate in 20% glycerol for 30–60 min.
(iii) Freeze-thaw four to six times by placing coverslips covered in glycerol into liquid nitrogen and thawing at room temperature.
(iv) Wash coverslips in 3 × 5 min in PBS.
(v)   Incubate coverslips in 0.1 M HCl for 5 min.

Or (b) plant tissue sections:
Apply 15 µl of enzyme solution to each well and incubate for 45 min.
6. Wash preparations in 2 × 10 min in 2× SSC.

Notes

1. We recommend preparation of fresh fixative from paraformaldehyde powder that has not been in contact with air, immediately before use. Instead of 1× PBS, 75 mM phosphate buffer (Appendix 1) can be used.
2. Many plant tissues, due to thick cell walls particularly of epidermal cells, show poor penetration of fixative. Adding detergents to the fixative will help penetration: 0.1% Tween 20 and 0.1% Triton X-100 is recommended.

3. Time of fixation is critical and needs adjusting to the material. Underfixation may result in poor morphology and loss of nucleic acid, while over-fixation masks target sequences.
4. See *Protocol 5.3* for more detail on enzyme solutions.
5. Protocols for handling cells grown on coverslips supplied by Irena Solovei and Thomas Cremer, University of Munich, Germany. Plant tissue preparation is based on Rawlins *et al.* (1991).

Additional protocols for whole mounts of insect and vertebrate embryos for RNA in situ hybridization

Whole mounts of various insect and vertebrate material, in particular immature embryos of *Drosophila*, grasshopper, *Xenopus*, chicken and mouse have been used to demonstrate gene expression during development and organogenesis (Tautz and Pfeiffle 1989; see Wilkinson 1992). Either the whole embryo or thick sections are processed. The protocols use the same fixation as given in *Protocol 6.8*. It is important to open up any cavities that might trap reagents prior to fixation. Pretreatment to remove membranes (e.g. vitellin of *Drosophila* eggs, see Tautz *et al.* 1992) or remove pigment granules (e.g. bleaching, see Wilkinson 1992) or endogenous phosphates might be necessary.

6.7 Centrifuge preparations (Cytospin and Cytoblock)

Cells from a suspension can be deposited onto glass slides without the use of acetic acid fixation using centrifuge devices (*Protocol 6.10*). The most commonly used centrifuge device for this purpose is the Cytospin (Shandon), although other manufacturers make fitments for centrifuges, or, since speeds used are low, devices can be made for old centrifuges. Cells can also be centrifuged with less force to embed in paraffin for sectioning (Cytoblocks). If analysis of chromosomes of tumor tissue is needed, but short-term culture of tumor cells is not possible, nuclei can be isolated from frozen material using pepsin and deposited on a slide using a centrifuge. A suitable protocol is given by Hopman *et al.* (1992).

| PROTOCOL 6.10 | PREPARATION OF CENTRIFUGE PRODUCTS (CYTOSPINS) FROM CELL SUSPENSIONS | 6.10 |

Reagents and equipment

Dulbecco's PBS pH 7.0 (e.g. Sigma D5773).
TBS (Appendix 1; made up RNase-free).
96% ethanol.
Paraformaldehyde fixative (see *Protocol 6.3*).

Centrifuge with slide adaptor and associated fittings: e.g. Shandon Cytospin™ 3.
Washed (*Protocol 5.1*) or coated (*Protocol 6.2*) microscope slides.

Method

1. Centrifuge cell suspensions at 100g for 5 min, decant the supernatant and resuspend the cells in 10 ml of PBS.
2. Repeat Step 1 twice.
3. Centrifuge and remove as much supernatant as possible then resuspend the cells in 1× RNase-free TBS to give a concentration of 2 × 10^6 cells ml^{-1}
4. Prepare centrifuge slide assemblies following manufacturers instructions using washed and coated microscope slides.
5. Add 100 μl of cell preparation into each funnel in the slide apparatus. Centrifuge for 5 min at 1500 rpm with low acceleration (Note 1).
6. Remove slides from the assemblies and fix in paraformaldehyde fixative in a staining jar for 5 min.
7. Wash in RNase-free TBS for 2 × 5 min.
8. Dehydrate slides by immersion in 90% alcohol for 3 min then 96% alcohol for a further 3 min.
9. Allow slides to air-dry and store at room temperature protected from dust.

Notes

1. Times and settings are given for a Shandon Cytospin™ 3 and might need adjustment for other models.

2. Protocol supplied by Tony Warford, Cambridge Antibody Technology, Cambridge.

Additional protocol for the preparation of paraffin wax blocks from cell suspensions

Protocols follow methods supplied with the centrifuge apparatus. Using a Cyto-block (Shandon) system, after paraformaldehyde fixation and washing (see *Protocol 6.9*), cells are mixed with Cytoblock Reagent 2 and transferred to a Cytoblock cassette before spinning at 500 rpm with low acceleration. Open the lid of the cassette and add one drop of Cytoblock Reagent 1 to the well. Close the lid of the cassette and place in 70% alcohol. The block cassette is processed overnight using a routine paraffin wax schedule (see *Protocol 6.3* or its additional protocol).

CHAPTER 7

Stringency and kinetics

The homology between the probe (labeled nucleic acid) and target (e.g. chromosomal DNA or cellular RNA) that is required for a hybrid (double helix) molecule to form and remain stable during *in situ* hybridization must be decided before the experiment. This information is essential for interpretation of results, although often not presented in manuscripts. Stringency is calculated using the melting temperature, T_m, of the DNA or RNA: the temperature at which double-stranded and single-stranded molecules are equally stable.

The stringency defines the percentages of matches and mismatches between probe and target nucleic acids that are allowed to occur without the double helix hybrid falling apart. At 85% stringency, a hybrid with 85% bases or more along the probe–target hybrid being complementary (homologous), and 15% or less mismatched, will remain stable; hybrid molecules with 80% homology will not form or dissociate immediately. At the T_m, the stringency is 100% and the probe and target will remain hybridized only where they are perfectly matched at all base pairs. At stringencies calculated above 100%, probe–target hybrids cannot form – the nucleic acid becomes denatured, and such conditions are used to make the probe and target single-stranded before the experiment. In practice, hybridization will occur between random probe and target sequences at about 40% stringency because of the base composition of DNA.

7.1 Choice of stringency

In situ hybridization is typically carried out in the range of 70% (described as 'low' stringency) to 90% ('high' stringency). Stringencies outside this range may be used, for example lower to localize heterologous probes on a target from a different species, or higher to localize chromosome-specific variants of a sequence family. In practice, the lowest background signal is achieved by hybridizing the probe and target at one stringency, and then carrying out post-hybridization washes at a stringency 2–5% higher. Weak probe–target complexes which have formed during the overnight hybridization denature and unhybridized probe is washed away.

7.2 Factors affecting stringency

The factors affecting stringency have been determined experimentally for DNA in solution, and theoretical principles and parameters incorporated into formulae. Because single-stranded and hybrid molecules have different light absorption at 260 nm, continuous spectrophotometry in temperature-controlled cells can be used to assay dissociation and reassociation accurately. Extension of the formulae for use in both Southern (membrane) and *in situ* hybridization are empirical. Results in both human and plants using variants of sequences as probes (e.g. in plants: Kamm *et al.* 1995; Schmidt amd Heslop-Harrison 1996b), target chromosomal sequences with known sequence variants (human: Schwarzacher *et al.* 1988), and different synthetic simple sequence repeat oligomers (Schmidt and

Heslop-Harrison 1996a), suggest that the stringencies calculated using the formulae are correct. However, we caution that several examples, mostly unpublished, are known where RFLP analysis by Southern hybridization does not correlate with the location (the order of three probes) or copy number (number of sites detected) determined by *in situ* hybridization. Some discrepancies can be explained, for example where RFLP exposures are too short to show all bands, or non-segregating, unmapped, bands are present, or the chromosomal sequence is not accessible to the probe. These seem not to explain all inconsistencies indicating that an unknown equation involving the factors below and perhaps a ratio between probe length, homologous target site length and probe–target homology may be different between solution, membrane and *in situ* hybridization.

Stringency is controlled by base pair composition of the probe, the nature and length of the probe fragments after labeling, temperature, and the composition of the hybridization solution, particularly the concentration of monovalent cations (in most cases sodium) and formamide. Temperature, sodium ion and formamide concentrations are varied to control stringency during *in situ* hybridization experiments. In summary, higher temperatures, higher formamide and lower salt concentrations give higher stringency (lower stability of the double helix). pH, probe availability and additional factors including the presence of Mg^{2+} or other ions have effects not taken into account by key formulae (see Section 7.3).

Temperature: Normally, temperatures substantially higher than the melting temperature are used to denature both probe and target DNA. During hybridization and washing, temperature must be controlled accurately to define the stringency of the experiment. A change of 1°C alters stringency by 1%. Critical steps are usually carried out at 37 or 42°C, but temperatures can be varied between 25 and 65°C (preparations may be lost from slides or morphology damaged at higher temperatures).

Monovalent cation concentration: At low ion concentrations, the double helix is less stable, while high concentrations make it more stable. The concentration of monovalent cations, almost always sodium, is controlled by the concentration of the weak buffer SSC (saline sodium citrate) or a sodium phosphate buffer, varying from more than 1 M to less than 20 mM (6× SSC and 0.1× SSC; see *Table 7.2*).

Formamide: This organic liquid (CH_3NO) destabilizes the double helix: DNA:DNA hybrids to a greater extent than RNA:RNA hybrids. Its inclusion in hybridization and wash solutions, often between 20 and 60%, allows lower temperatures to be used for high stringency experiments that are gentler to chromosome spreads and sections. Older protocols used high volumes of formamide, but it is toxic and volatile, so amounts are now reduced and it is often omitted from post-hybridization washing solutions.

Nature of probe and target: RNA:RNA hybrids are 10–15°C more stable than DNA:DNA hybrids; RNA:DNA hybrids are intermediate (Wetmer *et al.* 1981).

pH: Under both acid and alkaline conditions, DNA will denature. pH is not normally varied to control hybridization stringency and has little effect between 5 and 9, although high pH has been used occasionally to increase stringency. Chromosome preparations can be placed in either an acid or alkaline bath for denaturation. SSC, or a sodium phosphate buffer, with a pH of 7.0, are almost always included in the hybridization mixture to provide buffering.

Probe fragment length: This refers to the length of probe fragments after labeling. Short probes form less stable hybrids with targets than long probes. Kinetics is also affected by probe fragment length (see Section 7.4). A shorter fragment with 100% homology to the target may be more stable than a long fragment with lower homology, although this depends on where the mismatches occur.

GC content of probe: G–C base pairs form with three hydrogen bonds, while A–T pairs form with two. Thus, stable probe–target combinations will form between GC-rich sequences under more extreme conditions (higher temperature, higher formamide, lower salt) than for AT-rich sequences. Because DNA sequences used as probes can have very different base-pair ratios, this consideration may limit use of simultaneous double target hybridization. GC content cannot be altered as part of an experiment. Mammals typically have a 42% GC content and angiosperm plants have about 45% GC content. These figures may drop below 35% for repetitive sequences (Heslop-Harrison *et al.* 1999), or rise to over 70% in some bacteria, while synthetic oligonucleotides with extreme values (100% or 0%) may be used in some experiments.

Mismatches: There is a complex relationship between probe state and T_m, not entirely reflected in the formulae in Section 7.3 for the conditions found with *in situ* hybridization. The label groups are not attached to atoms involved in hydrogen bonding, but their size is likely to affect helix stability, so excess label incorporation is probably undesirable. The distribution of nonmatching base pairs along the probe, particularly with respect to the 5′ end, will alter stability, so it is conceivable that a PCR-labeled probe (*Protocol 4.3*), where all molecules are the same, will give different results from random prime labeling (*Protocol 4.2*) where each probe molecule has a different end. 'Tails' to the probe (non-homologous stretches at one end of the probe, obtained from a plasmid, or by end labeling of a sequence) may have little effect on the T_m of hybrid molecules, and the single-stranded tail may make the hybrid easier to detect.

7.3 Calculation of stringency

Stringency calculations have been worked out in detail based on short oligonucleotide hybridization and hybridization of longer DNA molecules in solution. Parameters for hybridization of probes in solution to chromosome preparations on a slide or membrane-bound DNA are empirical extensions, but data indicate that the theoretical values correlate well under most conditions with those achieved in practice.

The formula of Meinkoth and Wahl (1984) is used to calculate the melting temperature, T_m, of a DNA:DNA hybrid molecule for probes of 50 bp length or more:

T_m = 0.41 (%GC of probe) + 16.6 log (molarity of monovalent cations) − 500/(probe fragment length) − 0.61 (% formamide) + 81.5°C.

Some authors (Boehringer Mannheim, 1996) ignore the contribution of cation concentration above 0.4 M (i.e. 2× SSC) and use a factor of 0.72 (% formamide).

The melting temperature for RNA is given by Wilkinson (1992):

T_m = 0.58 (% GC of probe) + 0.0012 (%GC)^2 − 18.5 log (molarity of monovalent cations) – 820/(probe fragment length) − 0.35 (% formamide) + 79.8.

Typically, values for RNA:RNA melting temperatures are 10–15°C higher than the equivalent for DNA:DNA hybrids (Cox *et al.* 1984).

For both DNA and RNA hybrids, the melting temperature is converted to stringency (%) by the formula:

Stringency = 100 − $M_f(T_m − T_a)$

where T_a = actual temperature of hybridization or washing and M_f = mismatch factor, a linear relationship between 1 for probes >150 bp and 5 for <20 bp.

Table 7.1 gives melting temperatures for DNA:DNA hybrids at various SSC and formamide concentrations based on a GC content of 43%. For each 5% increase in GC content of the probe, add 2°C (5% decrease, subtract 2°C).

Table 7.1 Melting temperatures for DNA:DNA *in situ* hybridization calculated for probes with 43% GC content and 300 bp length using the formula in Section 7.3

SSC	Formamide (%)												
	60	55	50	45	40	35	30	25	20	15	10	5	0
5×	60.7	63.7	66.8	69.8	72.9	75.9	79.0	82.0	85.1	88.1	91.2	94.2	97.3
2×	54.1	57.1	60.2	63.2	66.3	69.3	72.4	75.4	78.5	81.5	84.6	87.6	90.7
1×	49.1	52.1	55.2	58.2	61.3	64.3	67.4	70.4	73.5	76.5	79.6	82.6	85.7
0.75×	47.0	50.1	53.1	56.2	59.2	62.3	65.3	68.4	71.4	74.5	77.5	80.6	83.6
0.5×	44.1	47.1	50.2	53.2	56.3	59.3	62.4	65.4	68.5	71.5	74.6	77.6	80.7
0.2×	37.5	40.5	43.6	46.6	49.7	52.7	55.8	58.8	61.9	64.9	68.0	71.0	74.1
0.1×	32.5	35.5	38.6	41.6	44.7	47.7	50.8	53.8	56.9	59.9	63.0	66.0	69.1

Table 7.2 gives the stringency of DNA:DNA hybridization and washing at the two most frequently used temperatures, 37 and 42°C.

As a rule-of-thumb, a 1.5% decrease in formamide concentration increases the melting temperature by 1°C, and a 1°C increase in temperature increases stringency by 1%.

The parameters for oligonucleotide hybridization (typically between 11 and 22 nucleotides long) are slightly different to those for longer probe fragments, and have been calculated by Wallace *et al.* (1981) for a monovalent cation concentration of 0.9 M:

$$T_m = 2 (A+T) + 4 (G+C) \text{ °C.}$$

PCR primer design programs (see *Protocol 4.3*) can be used to calculate more exact values. A hybridization temperature of 5°C below the T_m is usually chosen for *in situ* hybridization of perfectly matched sequences, but for every mismatch, a further 5°C decrease in temperature is required to maintain hybrid stability (Durrant and Chadwick 1994).

7.4 Probe concentration and kinetics

The rate of hybridization of a probe and target depends on probe concentration and fragment length. The hybridization rate varies somewhat with temperature,

Table 7.2 Percent stringency of DNA:DNA *in situ* hybridization for probes with 43% GC content and 300 bp long at temperatures commonly used. Values used in the protocols of this book are in bold

SSC	Formamide (%)												
	60	55	50	45	40	35	30	25	20	15	10	5	0
(A) At 37°C													
5×	76	73	70	67	64	61	58	55	52	49	46	43	40
2×	83	80	**77**	74	71	68	65	62	59	55	52	49	46
1×	88	85	82	79	**76**	73	70	67	64	60	57	54	51
0.75×	90	87	84	81	78	75	72	69	66	63	59	56	53
0.5×	93	90	87	84	81	78	75	72	69	65	62	59	56
0.2×	100	96	93	90	87	84	81	78	75	72	69	66	63
0.1×	105	101	98	95	92	89	86	83	80	77	74	71	68
(B) At 42°C													
5×	81	78	75	72	69	66	63	60	57	54	51	48	45
2×	88	85	**82**	79	76	73	70	67	64	60	57	54	51
1×	93	90	87	84	**81**	78	75	72	69	65	62	59	56
0.75×	95	92	89	86	83	80	77	74	71	68	64	61	58
0.5×	98	95	92	89	86	83	80	77	74	70	67	64	61
0.2×	105	101	98	95	92	89	86	83	**80**	77	74	71	68
0.1×	110	106	103	100	97	94	91	88	85	**82**	79	76	73

although preservation of chromosome morphology and control of stringency are more important reasons to choose hybridization temperatures.

Hybridization rate is proportional to the square of the probe concentration, so a higher concentration hybridizes much faster. The hybridization solution normally includes one or more hydrated components such as dextran sulfate, polyvinyl pyrrolidone or Ficoll to increase the effective probe concentration, giving a large enough volume of solution to cover the preparation without using excessive probe. The probe may take a long time to 'find' targets where there are few copies per cell, or one copy per nucleus. Parameters affecting re-annealing of DNA in solution have been worked out using concentration–time (Cot) curves, and shown the presence of different fractions of nuclear DNA which differ in the time taken to re-anneal. For the fastest fraction, consisting of highly repetitive sequences (Cot = 1), a few hours or even minutes (see Section 3.1 and e.g. in the case of highly repetitive probes or oligonucleotides for primer extension, *Protocol 8.4*) may be sufficient for *in situ* hybridization, while times of 3 days or more may be helpful for hybridization to single-copy targets.

Theoretical considerations would suggest that probe concentration, once it is in great excess of the target concentration (the case for almost all *in situ* hybridization experiments with cloned probes), will have little effect on hybridization, but experimental data indicates this is not the case and too much probe creates background. A four-fold increase in probe concentration already in great excess of the target may give interpretable signal where none was visible before, or very high and non-interpretable 'background' signal where clear signal was seen before. In the case of genomic DNA used as a probe, the amount of probe DNA and DNA target on the slide may be similar (e.g. 100 ng of wheat DNA is present on a typical metaphase preparation with 2000 nuclei, and 100 ng of probe might be used for hybridization). When probed to slides with hybrid nuclei, hybridization may be specific to one genome leaving the other genome with little label, even although the non-target genome probed alone on control slides may show relatively strong signal. If introgression of small alien chromosome segments is being examined using total genomic DNA from the alien species, genomic probe concentration may need adjustment to optimize specificity of the hybridization.

With oligonucleotide and longer probe fragments, hybridization probably occurs by random diffusion events matching a few base pairs, with low stability, and then the double-helix extending from this nucleation where probe and target are complementary. In solution the hybridization rate is proportional to the square root of the probe fragment length (longer is faster). However, under the conditions of *in situ* hybridization, the faster diffusion rate of shorter probes and their better penetration into the cellular or chromosomal matrix means the effect is not seen clearly. Experimental data indicate that DNA: DNA hybridization with probes longer than 1 kb is poor.

As well as probe–target hybridization, probe–blocking DNA and probe–probe hybridization can occur. Unlabeled, blocking, DNA is included in the hybridization solution to out-compete probe binding to non-specific nucleic acid binding sites, often proteins, present in the preparation. However, the effective probe concentration can be reduced if probe also hybridizes with the block. In some cases, unlabeled blocking DNA may be included in the hybridization mixture to prevent hybridization of probe to particular sequences or classes of sequence in the target, such as repeat sequences present in a long clone, or sequence common between two genomes in a genomic DNA probe. Both block–target and block–probe hybridization occur, and prevent the undesired probe–target combinations occurring, for example in genomic *in situ* hybridization (Section 3.1). Some protocols pre-anneal either block and probe, or block and target (see CISS, Section 3.1). Probe–probe hybridization (renaturation) occurs where both strands of probe are present but is undesirable since it removes single-stranded probe. It is important that probe does not re-anneal before addition to the preparation, particularly when there is no combined renaturation step as in alternatives to *Protocol 9.2*. Synthetic oligonucleotide probes, riboprobes (*Protocol 4.5*) and random primer labeling (*Protocol 4.2*) of single-stranded DNA probes (e.g. single-stranded M13 vectors) do not re-anneal, but only hybridize to one strand of a chromosomal DNA target.

CHAPTER 8

DNA:DNA *in situ* hybridization

Procedures for *in situ* hybridization of labeled DNA probes to spread or sectioned chromosomes and nuclei (DNA) are described here. Double-target experiments are recommended for most applications, while multiple-target or reprobing of preparations are possible if more information is required. Specialized alternatives use cells in suspension or whole mount preparations as targets for hybridization (Section 8.6). Detection and visualization of probe–target hybrids is described in Chapter 9 and procedures to detect RNA targets are covered in Chapter 10. Most experiments use long probes (from clones or genomic DNA, on average 100–500 bp long after labeling). Labeled synthetic oligonucleotides (14–30 bp long) are being used increasingly as probes, when salt and temperature conditions need to be adjusted. Oligonucleotides are also used for *in situ* primer extension and *in situ* polymerase chain reaction (PCR) protocols, particularly to detect repetitive sequences (*Protocol 8.5*).

THE NORMAL DNA:DNA IN SITU *HYBRIDIZATION PROCEDURE*

1. Use of one or more tested, labeled probes (Chapter 4).
2. Use of checked slide or grid preparations carrying target sequences (Chapters 5 and 6).
3. Pretreatment of the chromosome preparation prior to the hybridization to remove RNA, cytoplasm and other cellular material, and to make the cells and chromatin permeable to the probe (*Protocol 8.1*).
4. Refixation of the material to help maintain chromosomal morphology, and prevent loss of material during subsequent steps (*Protocol 8.1*).
5. Decision on hybridization stringencies (Chapter 7) and making of hybridization mixture (*Protocol 8.2* for cloned or genomic probes, *Protocol 8.3* for oligonucleotides).
6. Denaturation (making DNA single-stranded) of probe and chromosomes, before hybridization, normally overnight, when probe–target hybrids will form (*Protocol 8.4*).
7. Stringent washes to remove unbound or loosely bound DNA probe (*Protocol 9.1*).
8. Detection of hybridization sites and counterstaining of the chromosomes (*Protocols 9.2* and *9.3*).

During the *in situ* hybridization procedure it is important to work accurately and cleanly to ensure results are reproducible, to reduce the accumulation of dirt that causes background, and to reduce loss or damage to the material. All tools and containers must be clean, and solutions without fungal or bacterial growth, although aseptic conditions are not necessary. The preparations on slides must not touch other slides (a frequent cause of scratching material), nor dry out or accumulate water when incubating and between steps. Wash solutions should cover slides totally and should be changed carefully to avoid strong turbulence. In the average experiment, it is convenient to handle eight slides at once: these fit easily into one staining jar for the washing steps. Fifteen or 20 slides can be handled if two staining jars are used or extra slides are placed diagonally in the jars (but it is easy to leave slides behind in the washes).

Most protocols involve an overnight hybridization step of about 16 h, sufficient time for most probes to find homologous sequences on the chromosomes. It also gives ample time on the first day to set up the experiment; the next day has enough time for the stringent washes and detection. Several one-day protocols are published using only 2–4 h of hybridization; they work well for repeated sequences, and primer-extension protocols require even less primer annealing time (*Protocol 8.5*).

8.1 Special equipment

Equipment required for multiple protocols in Chapters 8–10 is shown in the box below

EQUIPMENT FOR IN SITU *HYBRIDIZATION*

1. *Plastic coverslips*: approximately 25 × 30 mm and 22 × 22 mm, cut from plastic autoclavable waste disposal bags (e.g. Sterilin) or 'cook-in-the-bag' oven bags available in supermarkets. Up to 400 μl of fluid can be trapped underneath the larger size, ideal for pretreatment and detection steps, and bubbles and coverslips can be removed easily without scratching the sample preparation. We use the 22 × 22 mm size for the hybridization step with 30–50 μl of hybridization mixture. Some researchers prefer siliconized or silanized glass coverslips (e.g. Sigma C0465) that need only 15–20 μl of hybridization mixture for a 22 × 22 mm area.

2. *Plastic (or glass) staining jars* (holding slides vertically; e.g. Azlon or Sigma); typical staining jars hold eight slides in slots and 100 ml of solution. Glass jars will tend to break in water above 50°C, although are easier to use since they are more stable in water baths.

3. *Humid chamber*. Any covered container that will hold paper towels soaked with water or buffer, and two rods as rests to hold the glass slides horizon-

tally. It is important that the lid is sloped to prevent condensed water dropping onto the incubating slides.

4. *Incubator or water bath at 37°C* (sometimes 42°C).

5. *Digital thermometer* with external probe (range −20°C to 100°C, cost *c.* $30) is very convenient; glass thermometers can be used. Regularly and consistently, check temperatures of solutions (placing the probe carefully between slides without scratching them in the liquid), denaturation plates, ovens etc.

6. *Programmable temperature-controlled heating block* (normally based on a PCR machine; e.g. Hybaid, Techne or others; see Heslop-Harrison *et al.* 1991); slide or photographic dish warmers can be used but are less accurate and may not be hot enough). A 90°C water bath can be used (see *Protocol 8.4*, Step 2, and *Protocol 8.3*).

7. *Water bath* (with variable temperature and preferably shaking) for denaturation of the probe (60–95°C) and post-hybridization washes (35–50°C).

8.2 Pretreatment of chromosome preparations for DNA:DNA *in situ* hybridization

Pretreatments of slide preparations, whether chromosome spreads, sections or whole mounts, are required for three purposes:

(1) to remove extraneous RNA and proteins which will bind to probe and detection reagents, increasing background;

(2) to enable access of probes to the DNA – permeabilizing the target material;

(3) to fix the preparation so chromosomes and nuclei are not lost from the slide during the procedure.

PROTOCOL 8.1 PRETREATMENT OF CHROMOSOME PREPARATIONS

Reagents

2× SSC: dilute from 20× SSC stock (Appendix 1).

RNase solution: DNase-free ribonuclease A (e.g. Sigma R4642, solution in 10 mM Tris-HCl, pH 8 and 50% glycerol, 70 units mg protein^{-1}). Make up stock of 10 mg ml^{-1} in 10 mM Tris-HCl, pH 8 (Appendix 1). Store at −20°C in aliquots. Dilute to 100 μg ml^{-1} prior to use (Note 1).

10 mM HCl.

Pepsin (from porcine stomach mucosa, e.g. Sigma P6887, 3200–4500 units mg protein^{-1}). Make up stock of 500 μg ml^{-1} in 10 mM HCl and store in aliquots at −20°C. Dilute with 10 mM HCl to 1–10 mg ml^{-1} prior to use (Note 2).

Paraformaldehyde fixative. In the fume hood, add 4 g of paraformaldehyde (EM grade) to 80 ml water, heat to 60°C for about 10 min, clear the solution with a few drops of concentrated (10 M) NaOH, let cool down and adjust the final volume to 100 ml with water. Adjust pH to 8 with 1 N H_2SO_4 (the solution has no buffering capacity so changes pH rapidly). The paraformaldehyde solution is best prepared freshly (during the RNase and pepsin incubations), but can be stored for a few days at 4°C.

Ethanol: 96%, 90% and 70% in water.

Alcohol: acetic acid fixative (optional).

Material

Good quality selected slides from preparation protocols (spreads, cytospin, whole mounts and sections, Chapters 5 and 6). Check using phase-contrast microscopy for amount of remaining cytoplasm (to decide time and concentration of any pepsin treatment needed), write numbers on slides, and mark the area of interest on each slide by scratching underneath with a diamond pen. Pencil writing on frosted-end slides, but not marker pens, will survive the washes and temperatures. It can be very difficult to identify the sample side and area as slides are placed into and removed from washes, but the scratches can be seen and felt easily. The time since fixation for material used for the preparation, and the storage conditions of the slide preparations, affect the length of pretreatments required (less for young fixations; Note 2), as well as denaturation times and temperatures. We recommend keeping slides overnight at 37°C after preparation, and then storing at −20°C for longer (up to several months). Some authors report better results from slides stored for 2 weeks at room temperature.

Method

For incubations on the slide, an ample volume, typically 200 μl, is applied to the sample area and covered with a large plastic coverslip to avoid drying out. For the washing steps, slides are put into staining jars and covered totally with solution (typically 80–100 ml for up to eight slides). Agitation by gently moving the staining jars by hand (once every minute) or continuously on a shaking platform is recommended. To change solutions, pour off carefully and replace slowly with the next solution, or transfer slides to another staining jar with the next solution. Do not let slides dry out between changes of solutions.

Carry out steps at room temperature if not otherwise stated.

1. The slides may be re-fixed and cleaned, particularly if stored for more than a week, by putting into alcohol:acetic acid fixative for 10 min, washing twice in 96% ethanol for 10 min each and air-drying.

2. Add 200 μl RNase to the marked area of each slide, cover with a large plastic coverslip and incubate for 1 h at 37°C in a humid chamber (Note 3). Start preparing paraformaldehyde fixative.

3. Remove the coverslips carefully and wash slides in 2× SSC twice for 5 min.

4. Incubate slides in 10 mM HCl for 5 min. Take each slide, shake off excess fluid and quickly add 200 μl of pepsin, cover with a plastic coverslip and incubate for 10–15 min at 37°C in a humid chamber (Note 2).

5. Remove the coverslips and stop the reaction by placing the slides in distilled water for 1 min. Wash slides in 2× SSC twice for 5 min.
6. In the fume hood, place the slides in paraformaldehyde fixative for 10 min.
7. Wash in 2× SSC twice for 5 min.
8. Dehydrate slides through an ethanol series (70%, 90% and 96% ethanol, 2 min each). Air-dry in a rack. Do not dehydrate and dry sections or whole mount preparations.
9. Check dry slides by phase-contrast microscopy. Cytoplasm should be removed by the pepsin treatment, but chromosomes should not be lost or damaged (Note 4).
10. Proceed to denaturation and hybridization (*Protocol 8.4*) or *in situ* primer extension (*Protocol 8.5*). If necessary, slides can be kept overnight at 4°C.

Notes

1. The RNase must be DNase-free. If RNase is not purchased DNase-free, inactivate DNase in the RNase by placing the stock solution in boiling water for 15 min.
2. Pepsin should be used to help remove excess cytoplasm covering chromosome preparations. Adjust concentration to suit the preparations; if there is little cytoplasm, no pepsin treatment is needed. Most spread preparations are treated with 1 μg ml^{-1} for 10 min, but up to 10 μg ml^{-1} for 1 h can be used; however, it is often better to improve the preparation technique (see *Protocols 5.3* and *5.5*, e.g. the use of additional 3: 1 alcohol: acetic acid treatments or 60% acetic acid). For sections or whole mounts, proteinase K is more

effective to remove cytoplasm (see alternative protocol below), and can also be used with spreads.
3. RNase treatment is not essential for all materials and probes: RNA may be degraded already, and is not required for non-transcribed targets without RNA homology. Step 3 (acid wash) also degrades extraneous RNA.
4. If loss of cells and chromosomes has occurred, it suggests that the slides were not acid-washed (*Protocol 5.1*) before spreading of chromosomes, and most likely further loss will occur. It is better to stop the procedure and start with new preparations.

Alternative protocol to pretreat sections and whole mounts with proteinase K

Sections of material that have been fixed by cross-linking proteins with para-formaldehyde or glutaraldehyde need stronger enzyme digestions to remove proteins and to unmask DNA. This process of permeabilization increases accessibility of probe and detection reagents; proteinase K is recommended and its concentration and incubation time need to be adjusted carefully to suit the material. Proteinase K treatment can be used for squashed material that has been fixed in alcohol:acetic acid for several months or has cytoplasm that is difficult to remove (e.g. some meiotic cells), but even slight over-treatment will damage or lose chromosomes.

Instead of the pepsin solution of *Protocol 8.1*, prepare a stock solution of 500 µg ml^{-1} (e.g. Sigma P6556, 10–20 units mg^{-1}) in proteinase K reaction buffer (20 mM Tris-HCl pH 8, 2 mM CaCl$_2$); store at $-20°$C. Dilute with reaction buffer to 1–10 µg ml^{-1} prior to use. In Step 4 of *Protocol 8.1*, incubate slides in proteinase K reaction buffer for 5 min followed by incubation in proteinase K for 10 min at 37°C and stop digestions in reaction buffer containing 50 mM MgCl$_2$ (Step 5).

8.3 Hybridization mixture

The stringency at which an *in situ* hybridization experiment is carried out determines the approximate percentage of nucleotides that are correctly matched in the probe and target (Chapter 7, *Tables 7.1* and *7.2*). The stringency (in per cent) can be calculated by subtracting the difference in degrees centigrade between the actual hybridization temperature and the melting temperature of the probe DNA from 100, and is influenced by the GC content and lengths of the probe DNA, the salt and formamide concentration and the temperature of the reaction. Routinely (see *Table 8.1*), we use a final concentration of 50% formamide and 2× SSC (or 40% formamide and 1× SSC) at 37°C for hybridization (75–80% stringency) and 20% formamide and 0.1× SSC at 42°C for washing (80–85% stringency). If a probe gives unspecific dot-like signal to many chromosomal sites, and the background is not due to low quality chromosome preparation, it is worth increasing stringency (see also Chapter 12). If signal is weak, reducing the stringency is usually worthwhile. We normally change the formamide or SSC concentration rather than hybridization and washing temperature. *Protocol 8.2* and *Table 8.1* describe preparation of the hybridization mixture for standard-cloned, PCR-amplified or genomic-labeled probes. *Protocol 8.3* is for synthetic oligonucleotides such as simple sequence repeats, using a different hybridization mixture because of the different hybridization properties of short oligonucleotides, possibly with extreme base-pair composition (see Section 7.2).

It is often difficult to assess accurately the concentration of small amounts of DNA, particularly after labeling where it is not feasible to use a large part of the labeled yield for testing. The best test is an *in situ* hybridization run. It is not helpful to use excess probe DNA and too much probe will cause more unspecific signal on chromosomes and background, apart from being expensive. For genomic *in situ* hybridization (GISH), chromosomal *in situ* suppression (CISS) or use of clones including repetitive DNA, the ratio of amounts of probe to unlabeled genomic (or re-annealed Cot = 1) blocking DNA is more critical than absolute amounts. As a guideline, make the assumption that the conditions given for nick translation (*Protocol 4.1*) and random primer labeling (*Protocol 4.2*) yield 1 µg labeled probe. After PCR labeling (*Protocol 4.3*), with a reaction volume of 50 µl, test 5 µl on a gel, ethanol precipitate (*Protocol 4.6*) the rest and resuspend it in 25–50 µl according to strength of the band on the gel. Normally 1–2 µl will give a good *in situ* hybridization signal.

An inert DNA, usually isolated from salmon or herring sperm or *Escherichia coli*, is added to the hybridization mixture to out-compete binding of the probe to

Table 8.1 Hybridization mixtures. Two examples, each using two probes, are given, one for cloned repetitive sequences and one for total genomic DNA (see *Protocol 8.2* for further details)

Solution	Final concentration suggested	Examples of hybridization mixtures			
		Cloned sequence		Total genomic DNA	
		Final concn	Amount	Final concn	Amount
100% formamide (Note 1)	40–50%	50%	15 µl	40%	16 µl
20× SSC (Note 1)	1–2×	2×	3 µl	1×	2 µl
50% dextran sulfate	10%	10%	6 µl	10%	8 µl
10% sodium dodecyl sulfate	0.1–0.2%	0.165%	0.5 µl	0.125%	0.5 µl
Salmon sperm DNA 1 µg µl^{-1}	1–5 µg slide^{-1}	1 µg	1 µl	1 µg	1 µl
Probe 1 (Note 2)	25–200 ng slide^{-1}	75 ng	1.5 µl	100 ng	2 µl
Probe 2 (Note 2)	25–200 ng slide^{-1}	75 ng	1.5 µl	100 ng	2 µl
Blocking DNA (Note 3)	2–100× probe			5 µg	5 µl
Water	As required		1.5 µl		3.5 µl
Total volume			30 µl		40 µl

Notes to Table 8.1

1. The formamide SSC concentration and temperature determine the stringency of the hybridization (Chapter 7). We normally use final concentrations of 50% formamide and 2× SSC, which allows sequences of 75–80% homology to form duplexes. For genomic *in situ* hybridization, we often find that 40 µl hybridization solution is too little to fit all components with 50% formamide and 2× SSC, so use 40% formamide and 1× SSC which also has a stringency of 75–80%.
2. For differentiation, each probe needs to carry a different label, e.g. biotin, digoxigenin or a fluorophore. Examples shown use two cloned probes or two genomic probes. More or only one probe may be used, and cloned and genomic probes can also be combined. Probe concentration is approximate.
3. Excess amounts of unlabeled blocking DNA (genomic DNA or re-annealed Cot = 1 DNA) are used to out-compete hybridization of probe to undesired target sequences (see Section 3.1, genomic DNA and Section 7.3). Appropriate multiples of probe concentration to try are: CISS for human chromosomes, 2× probe; blocking clones, 10× probe; genomic DNA probe for detection of genomes in plant hybrids, 50×. If genomic probes are combined with cloned probes, it is often necessary to increase the cloned probe concentration or reduce blocking DNA so hybridization of the cloned probe is not affected.

proteins or other nonspecific DNA binding sites present in the preparation. Dextran sulfate is included in the hybridization mixture to increase the volume without diluting the probe and sodium dodecyl sulfate (SDS) helps the penetration of the probe.

PROTOCOL 8.2 HYBRIDIZATION MIXTURE FOR CLONED AND GENOMIC PROBES

Reagents

Formamide: good but not the highest grade (e.g. Sigma catalogue number F7508). Immediately after purchase, store at −20°C in small aliquots for hybridization mixture preparation (Note 1). Formamide is a potential carcinogen.

20× SSC (Appendix 1): filter (0.22 μm) sterilize if it has not been autoclaved.

50% (v/v) dextran sulfate in water. Dissolve by heating to 65°C. Filter (0.22 μm) sterilize and store for up to a week at room temperature, longer at −20°C (Note 2).

10% (w/v) SDS (sodium dodecyl sulfate, also called sodium lauryl sulfate) in water, filter sterilize. Store at room temperature.

Sonicated or autoclaved DNA from salmon or herring sperm or *E. coli* (1 μg μl^{-1} treat DNA following *Protocol 3.2*, or purchase commercially). Fragments should be 100–300 bp long. Store in aliquots at −20°C.

Labeled and tested probe DNA (Chapter 4; typical experiments use 0.5–1 μl of resuspended probe per slide). Store at −20°C.

Optionally, unlabeled autoclaved blocking DNA (*Protocol 3.2*). Fragments should be 100–300 bp long. Store at −20°C.

Method

1. Calculate the content and quantity of hybridization mixture following *Table 8.1* and record them in *Form 8.1*. Typical hybridization mixture volume is 30 μl per slide, but 20–40 μl can be used depending on area of preparation on slide.
2. For each probe or probe combination, prepare hybridization mixture in a microcentrifuge tube according to your calculations on *Form 8.1* (Note 3).
3. Mix well and denature by floating the sealed tube in a water bath at 70°C for 10 min. Place tubes on ice for 5–15 min.
4. Proceed to *Protocol 8.4* for denaturation and hybridization.

Notes

1. Molecular biology grade formamide will have a pH around 7 and be solid at −20°C; it degrades at room temperature. Very high quality formamide can be purchased frozen, but for the protocols here, freezing immediately on receipt is adequate if fresh from the supplier. Formerly, an ion exchange resin (e.g. Amberlite, Merck) was added to purify crude grades not specified for molecular biology.
2. The solution is very difficult, but possible, to force through the filter. To pipet the dextran sulfate solution, cut off the end of the micropipet tip to give a larger orifice.
3. Nearly every slide will need a different mixture, depending on material and probe; however it is easier and often more accurate to make a 'master mix' of components common to all hybridization mixtures and partition this before adding individual components.
4. The hybridization mixture containing formamide was first described by Harper and Saunders (1981). The protocol described here is based on Schwarzacher *et al.* (1992a) and Heslop-Harrison *et al.* (1990).

Form 8.1: **In situ** *hybridization data sheet*

Run code: Name: Date:

Cross-reference to notes or photographs of slides or materials:

Component	Typical value (final concn) volume	1	2	3	4	5	6	7	8		
100% formamide	(50%) 15 μl										
20× saline sodium citrate	(2×) 3 μl										
50% dextran sulfate	(10%) 6 μl										
Salmon sperm DNA 1 μg μl^{-1}	(1 μg) 1 μl										
10% sodium dodecyl sulfate	(0.15%) 0.5 μl										
Probe 1*	25–200 ng										
Probe 2*	25–200 ng										
Probe 3*											
Probe 4*											
Blocking DNA*	2–100× probe										
Water	(to 30 μl)										
Total volume	30 μl										

Slide codes and details

* Insert details.

Denaturation times and temperatures for the hybridization mixture:

Denaturation times and temperatures for the chromosome preparations:

Hybridization time and temperature:

Washing conditions:

PROTOCOL 8.3 HYBRIDIZATION MIXTURE FOR OLIGONUCLEOTIDES

Reagents

20× SSPE: 3.6 M NaCl, 200 mM NaH_2PO_4, 20 mM EDTA, pH 7.4. Sterilize and store at 4°C.

100× Denhardt's solution: 20 g l^{-1} Ficoll, 20 g l^{-1} polyvinylpyrrolidone, 20 g l^{-1} BSA (bovine serum albumin) in water. Store in aliquots at −20°C (cannot be autoclaved).

10% (w/v) SDS in water, filter sterilize. Store at room temperature.

Sonicated or autoclaved DNA from *E. coli* (1 μg $μl^{-1}$; *Protocol 3.2*). Fragments should be 100–300 bp long. Store in aliquots at −20°C.

Labeled single-stranded probe DNA: typically end-labeled oligonucleotides (*Protocol 4.4*; typical experiments use 1 μl per slide from 25 μl resuspended from the labeling reaction).

Method

1. Prepare the hybridization mixture with a volume of 30 μl per slide:

7.5 μl	20× SSPE (final concentration 5×)
1.5 μl	100× Denhardt's solution (final concentration 5×)
3 μl	10% SDS (final concentration 0.5%)
1.5 μl	denatured *E. coli* DNA (final concentration 50 ng $μl^{-1}$)
X μl	1–2 pmol labeled probe DNA
Y μl	water to 30 μl
30μl =	Total volume

2. Proceed to denaturation and hybridization (*Protocol 8.4*) (Note 1).

Notes

1. If a single-stranded probe is used, no denaturation step is required. Denature other probes as in *Protocol 8.2*, Step 3.

2. This hybridization protocol is based on those used by Nanda *et al.* (1991), Schmidt and Heslop-Harrison (1996a) and Cuadrado and Schwarzacher (1999).

8.4 Denaturation and hybridization

To allow hybridization of the labeled probe DNA, the DNA target on the slide must be denatured to make it single-stranded. High temperatures in formamide/ salt solutions – often the hybridization mixture itself – are now used most frequently for denaturation (but see alternatives to *Protocol 8.4*). Denaturation conditions are changed by altering time and/or temperature, but are most likely to lie in the range stated in *Protocol 8.2*. The narrow window allowing successful denaturation without destroying chromosome morphology can only be worked out empirically and is judged both by the appearance of the chromosomes and the *in situ* hybridization signal. The chromosomes have to be treated long and hard enough to allow all the target sequences to become single-stranded: there is some DNA loss and swelling of the chromosomes. In adjusting time and temperature, we normally use 5 min at temperatures between 70 and 80°C, increasing time above 80°C. If the chromosomes have become ghost-like or look like inflated balloons, they are overdenatured. Underdenatured chromosomes might have perfect morphology, but show weak or no signal. This can also be caused by bad probe or detection reagents (see trouble-shooting Chapter 12). Because of differences in DNA packaging, interphase nuclei and metaphase chromosomes have slightly different denaturation requirements: metaphase chromosomes denature under less extreme conditions. We have found that chromosomes from different species, different tissues, after different times in fixation, made by different researchers and surrounded by more or less cytoplasm, need different denaturation conditions. Also denaturation might differ for the detection of tandemly repeated sequences (tending to need more extreme conditions than dispersed or single copy sequences).

For several years, we have used a modified thermal cycler for denaturation of chromosome preparations together with the hybridization mixture (Heslop-Harrison *et al.* 1991). It ensures accurate and reproducible temperature and time conditions during denaturation, as the thermal cycler adjusts to the environment of the slide using a temperature probe or calculation. This method is quick and also avoids the hazards of using large quantities of hot formamide otherwise required for slide denaturation. Because the conditions used are such that the target DNA is only just denatured (to preserve morphology), we suggest that the probe in the hybridization mixture is fully denatured under more extreme conditions before applying to the preparation.

PROTOCOL 8.4 DENATURATION AND HYBRIDIZATION

Materials

Single-stranded or denatured hybridization mixture from *Protocol 8.2* or *8.3*

Fixed and pretreated spread chromosome preparations from *Protocol 8.1*.

Method

1. Add appropriate hybridization mixture to each marked area on preparations. Cover with a small plastic coverslip. Make sure no bubbles are trapped underneath the coverslip. They can be easily freed by lifting the coverslip carefully (Note 1).
2. Heat slides to denature target on a heated block at 60–90°C, leave them for 5–10 min and then cool them down to 37°C (slow cooling is usually used) (Notes 2 and 3). Thermal cyclers have the most accurate temperature control, but a heated plate with good temperature control can be used. Alternatively, a water bath at 90°C with a lid (essential to maintain a high temperature) can be used. Slides are rested on glass rods lying on moist paper in a closed heat-resistant chamber floating in the water bath, with a digital thermometer probe inside the chamber next to the slides. The temperature of the chamber is equilibrated and the temperature within the chamber accurately controlled by opening and closing the lid of the water bath for denaturation.
3. Incubate slides at 37°C (42°C may be used to increase stringency) in a humid chamber, in the thermal cycler or on the heated block, typically for 16–20 h (Note 4). Check slides do not dry out or accumulate condensation.
4. Proceed to stringent washes (*Protocol 9.1*).

Notes

1. For hybridization times up to 20 h sealing of coverslips is not necessary. However, for hybridization times of several days it is recommended to use siliconized glass coverslips and to seal them with rubber cement to avoid drying out.
2. The range of time and temperature has to be adjusted to different species and material: 75°C for 5 min provides conditions that denature most DNA.
3. During denaturation, it is important that no condensation drips onto the slides or the hybridization mixture will be diluted and stringency decreased.
4. Hybridization time depends on the amount and length of probe and amount of target. For highly repetitive sequences, 2–4 h will be adequate, but it can be advantageous to leave single-copy sequences to hybridize for 72 h or more.
5. The protocol described here is based on Heslop-Harrison *et al.* (1991).

Alternative protocols for denaturation of chromosome preparations

Slides can be denatured in staining jars containing high volumes or denaturing solution at an appropriate temperature. In early experiments, acid or alkali treatments, or sometimes enzymes, were used, but now formamide–SSC mixtures are used most frequently. Typically, place 80 ml of 40% formamide, 1× SSC (final concentration) in a water bath with lid at about 90°C in a fume hood. Place a digital thermometer probe (or thermometer) behind a new slide in the jar, and allow the temperature to stabilize at 2°C above the denaturation temperature chosen, typically 72°C, by opening and closing the water bath lid. Place two slide preparations into the solution, which will cool by 2°C, and control the temperature by opening and closing the lid. After denaturation, dehydrate the slides in an ice-cooled ethanol series and air-dry or remove slides and shake off excess fluid. Apply the single-stranded or denatured hybridization mixture and coverslip for hybridization (*Protocol 8.4*, omitting Step 2). This denaturation method is advantageous for reprobing slides since the high volumes of solution will denature and remove ('strip') the probes from the preparation.

Additional protocol for hybridization of sectioned and whole mount preparations

Section and whole mount preparations should not be air-dried after pretreatments (see Step 8 of *Protocol 8.1*). Better penetration of hybridization mixture to chromosomal target and correct denaturation is achieved by soaking slides in a pre-hybridization solution containing 50% formamide and 2× SSC (or equivalent depending on hybridization mixture) for several hours to days before starting *Protocol 8.4* (see Scherthan *et al.* 1994; Visser *et al.* 1998). Slides are removed from the pre-hybridization mixture and excess solution is shaken off before applying the hybridization mixture (Step 1).

8.5 Primer extension *in situ* (PRINS)

In situ hybridization using thermostable polymerases to extend DNA from sites of oligonucleotide hybridization is an area of extensive recent research. Newly replicated DNA is detected by incorporation of labeled nucleotides. Protocols using single-stage extension (primed *in situ* hybridization, PRINS; Gosden *et al.* 1991; Koch *et al.* 1992), two-stages as presented in *Protocol 8.5*, and temperature cycling are available. However, it should be noted that, despite the attractiveness of the method, there are still problems with high background and false signals. Therefore the method tends to be used for a minority of special applications of *in situ* hybridization. Three books on the topic have been published in the late 1990s and give extensive details and protocols (Bagasra and Hansen 1997; Gosden 1997; Herrington and O'Leary 1998).

The protocol given here is particularly useful for quick detection of sites of tandemly repeated DNA sequences and localization of small variants of repeat units. After alcohol:acid fixation and spreading of chromosomes in acetic acid (*Protocols 5.3–5.5*), there are many breaks in the DNA, and these renature in the preparation so there are numerous self-priming sites which will give background and prevent use of low copy primers. To reduce self-priming, a first round extension (without added primer or label) can be carried out incorporating dideoxynucleotides to block further extension from self-priming sites (*Protocol 8.5*). A second round is then used for specific extension with addition of a synthetic primer and labeled base. Because of the possibility of self-priming and incorrect priming, it is very important to include multiple control slides: without added primer, with a primer known not to be present in the target chromosomes, and with an abundant primer.

PROTOCOL 8.5 PRIMER EXTENSION *IN SITU* (PRINS)

Reagents

Oligonucleotide primers including control primers: 14–22 nucleotides long with known annealing temperatures (see *Protocol 4.3* for primer design considerations).

Primer extension buffer: 5 mM Tris-HCl, pH 8.3 (Appendix 1), 25 mM KCl, 0.005% gelatine, 0.005% Tween 20, 0.005% Triton X-100, 0.05% Nonidet P40 and 0.75 mM MgCl$_2$.

Unlabeled nucleotides: 2 mM stocks of dCTP, dGTP, TTP and dATP in 100 mM Tris-HCl, pH 7.5 (Note 1).

(Optional) dideoxynucleoside-5'-triphosphate (ddNTP): any one of ddCTP, ddGTP or ddATP, 2 mM in 100 mM Tris-HCl, pH 7.5 (Note 1).

Labeled nucleotide: prepare the required mixture in 100 mM Tris-HCl, pH 7.5.

For digoxigenin: 0.2 mM digoxigenin-11-dUTP (Roche) and 0.4 mM TTP.

For biotin: 0.4 mM biotin-11 (or 16)-dUTP (e.g. Sigma, Roche) and 0.2 mM TTP.

For direct fluorophore labels e.g. 0.3 mM fluorescein-dUTP or rhodamine-dUTP (e.g. Amersham) and 0.3 mM TTP (see Table 4.1 for alternative labels).

Taq DNA polymerase (1–5 units μl^{-1}; this application does not involve a chain reaction so no PCR licence is required).

2× SSC: dilute from 20× SSC stock (Appendix 1).

Ethanol series: 70%, 90%, 96% ethanol in water.

Detection buffer: 4× SSC containing 0.2% (v/v) Tween 20.

Material

Fixed and pretreated chromosome preparations from *Protocol 8.1*.

Methods

For a two-round protocol with extension blocking, start at Step 1; for a single-round protocol, start at Step 4.

1. On each preparation, place a solution of 40 μl of primer extension buffer (no primer or labeled nucleotide) containing 1.5 units of *Taq* DNA polymerase and 2 μl of 1 mM stock of each nucleotide; one is ddNTP and the other three dNTPs.

2. After covering with a plastic coverslip, heat slides to 85°C for 8 min denaturation (or as appropriate for the material; see *Protocol 8.4*), leave at 37°C for 30 min of (self-) re-annealing, and then heat to 72°C for 30 min extension/termination in a modified thermal-cycling instrument or heated block.

3. Wash slides in 2× SSC and dehydrate through an ethanol series (5 min each before air-drying).

4. Place a solution of 40 μl of primer extension buffer containing 1.5 units *Taq* DNA polymerase, 5 ng primer, 1 μl of labeled nucleotide and 1 μl each of dCTP, dGTP and dATP onto each preparation and cover with a plastic coverslip (Note 2).

5. If single-round protocol is used, denature, typically at 85°C for 8 min. (Do not repeat denaturation if two-round protocol is used and preparations were denatured in Step 3.)

6. Allow primer to anneal at 5–10°C below the calculated annealing temperature, typically 55°C, for 40 min, followed by 72°C for 40 min for primer extension.

7. Slides can be left in a humid chamber at 4°C overnight at this stage.

8. Wash two times in detection buffer at 37°C, and two times in detection buffer at room temperature.

9. Proceed to detection if indirect labels are used (*Protocol 9.2*) or staining and mounting if fluorophores were used as label (*Protocol 9.3*).

Notes

1. Store in 100 μl aliquots for up to 3 months at −20°C. Even pure nucleotides degrade over 12 months at −20°C and should be replaced regularly. dNTP and ddNTP sets are available from e.g. Amersham, Roche, Sigma.
2. Essential controls include no primer, a primer not known to be homologous to the target and a primer known to be present (e.g. part of the rDNA sequence).
3. The method is based on Kipling *et al.* (1994) and Heslop-Harrison *et al.* (1999).

Alternative protocols for primed in situ *hybridization*

Many alternative protocols are given in the chapters of the three books discussed above. There are developments using temperature cycling for *in situ* PCR. Notable results have been obtained by Macas *et al.* (1995) using fragments of glass coverslips with chromosome preparations placed in PCR reaction solutions in microcentrifuge tubes, before temperature cycling.

8.6 Reprobing

Using fluorescent *in situ* hybridization and exploiting multicolor detection, several probes with different labels or label combinations can be used simultaneously. Often, though, it is desirable to reprobe a preparation after the first (or subsequent) hybridizations. If a low-copy probe is used, simultaneous use of a repetitive sequence can often obscure signal because of overlap of wavelengths, so it is desirable to localize repeats separately; probes may overlap in position so better resolution is obtained in separate experiments; or more data may be required after analysis of the first set of results. Reprobing may be also carried out at a different stringency. We have found that slides can be reprobed two or more times with only minor loss of chromosome morphology after both single and double target *in situ* hybridization (Heslop-Harrison *et al.* 1992), although whole chromosomes may occasionally be lost. We have found that no modification of denaturation times and temperatures is required for the second probing.

PROTOCOL 8.6 REPROBING OF PREPARATIONS

Reagents

Detection buffer: 4× SSC containing 0.2% (v/v) Tween 20.
2× SSC: dilute from 20× stock (Appendix 1).

(Optional) alcohol series: 70%, 90%, 96% ethanol in water.

Material

Slides after photography and analysis of the first probing results.
Single-stranded or denatured hybridization mixture

containing new labeled probe or probe combinations from *Protocol 8.2* or *8.3*.

Methods

1. Blot immersion oil off surface of coverslip and carefully wipe around coverslip with solvent to remove any traces of oil from the slide. Wipe underneath of slide.
2. Put slides in 37°C for 5–10 min to reduce the viscosity of the glycerol/antifade mountant. Remove coverslip by lifting slowly but steadily with the edge of a razor blade (Note 1).
3. Wash slides in a staining jar with detection buffer at room temperature for 5 min and then twice for 30–60 min (Note 2).
4. Incubate slides in 2× SSC twice for for 5 min at room temperature. Optionally dehydrate slides in an alcohol series two times each and then air-dry.
5. Apply hybridization mixture to the slide and redenature and hybridize as described in *Protocol 8.4* (Note 3).

Notes

1. Do not dip slides with immersion oil into solutions to remove coverslip; an oil film will form on the surface of the solution and adhere to the preparation when taking it out. It is not possible to wipe off oil from coverslips completely without the likelihood of moving the coverslip and damaging the preparation.
2. The 'washes' remove the antifade solution, but are unlikely to remove detection reagents and are not stringent enough to remove labeled probe. The denaturation stage will remove the labeled probe from the target, and the label will then be

diluted in the new hybridization solution. Nevertheless, residual signal from the first probing will almost always be visible (also see Note 3).
3. Denaturation by dipping the slide into a hot formamide solution (see *Protocol 8.4* alternatives) removes more of the first probe, which goes into the formamide solution when denatured. Then add the hybridization mixture, coverslip and allow to hybridize.
4. This protocol is based on Heslop-Harrison *et al.* (1992).

8.7 Solution hybridization

In situ hybridization to chromosomes in suspension in a solution has been used very successfully, particularly for applications in conjunction with flow cytometry, where chromosomes are labeled before sorting or after sorting to confirm their identity. Dudin *et al.* (1987) published protocols for human chromosomes, while Dolezel, Schubert and colleagues have carried out extensive work in plants (Macas *et al.* 1995; Pich *et al.* 1995).

CHAPTER 9

Washing and detection of DNA *in situ* hybridization to nuclei and chromosomes

After stringent washes to remove unbound and weakly hybridized probe (or directly after *in situ* primer extension, *Protocol 8.5*), various options are available for detection of probe hybridization sites (given in the flow diagram, *Figure 9.1*). In this chapter, we give protocols for fluorescent detection most frequently used with DNA targets. Colorimetric detection methods using enzyme-linked detection reagents are occasionally used, and are described in Chapter 10. The equipment required for *Protocols 9.1–9.4* is similar to that required for *in situ* hybridization (Chapter 8, except no high temperatures are required). Some manufacturers make special slide racks and heated chambers for transfer of slides between wash solutions; these may be faster than exchanging solutions or transferring individual slides between staining jars as suggested here but are not necessary.

9.1 Post-hybridization washes

The post-hybridization washes remove the hybridization mixture and unbound probe. Different protocols wash at slightly above or slightly below (but rarely more than 5% different from) the hybridization stringency. We suggest using washes that are a few percent more stringent than the hybridization to minimize background, as probe with weak but non-specific binding and some weakly hybridized probe is removed. Washes following hybridization with most cloned and genomic DNA probes will follow *Protocol 9.1*. An alternative protocol without formamide can be used, particularly with oligonucleotides. The absence of formamide is a safety and cost advantage for routine experiments and for teaching. To avoid extremes of temperature or very low salt concentrations, both of which may remove material from slides, non-formamide washes are normally carried out at lower stringencies than the hybridization.

It is critical that the temperature of the stringent washes is controlled accurately: an increase of a few degrees for only seconds may remove most of the probe, while reduced stringency washes will increase background signal. Temperatures should not exceed 45–60°C (depending on the nature of the preparation) to avoid loss of material.

PROTOCOL 9.1   STRINGENT WASHES

Reagents

2× and 0.1× SSC: dilute from 20× SSC stock (Appendix 1).

Formamide: good but not the highest grade. Immediately after purchase, store at −20°C in aliquots sufficient for the washes in one experiment, typically 40 ml (Note 1). Formamide is a potential carcinogen.

Stringent wash solution: typically, use a final concentration of 20% formamide and 0.1× SSC for high stringency (40 ml 100% formamide and 1 ml 20× SSC made up with water to 200 ml) or 2× SSC (40 ml 100% formamide and 20 ml 20× SSC made up to 200 ml) for low stringency (Note 2).

Method

For the washing steps, slides are put into staining jars and covered totally with solution. For Steps 4–7 use a gentle shaking water bath, rotating platform or hand-agitate the solution. To change solutions, pour liquid off carefully and replace slowly with next solution, avoiding turbulence, or transfer slides to another staining jar with the next solution. Do not let slides dry out between changes of solutions.

1. Prepare post-hybridization washing solutions. Typically, for eight slides in a 100 ml staining jar, you will need 500 ml 2× SSC, 200 ml 0.1× SSC and 200 ml stringent wash solution (the stringent wash solution should be prepared in a fume hood). Heat solutions to 42–45°C in a water bath.
2. Take slides from hybridization chamber, avoiding condensation dropping on to the slides. Examine slides and note anything unusual: Have any slides dried out? Has any water dropped on the slides? Are there any bubbles?
3. Float off coverslips by incubating slides in a staining jar of 2× SSC at 35–42°C. The plastic coverslips will fall away in the solution and are removed with a pair of forceps. Any surface film or dust must be wiped from the liquid in the jar with a tissue before pouring away solution. Otherwise it will stick strongly to preparations.
4. Wash with fresh 2× SSC at 42°C for 2 min.
5. Incubate slides in stringent wash solution twice for 5 min at 42°C (or temperature as required, Note 2). Measure the temperature of the solution in the staining jar, and remove from water bath or heat to maintain temperature to within 0.5°C (Note 3).
6. Wash slides in 0.1× SSC (high stringency) or 2× SSC (low stringency) twice for 5 min at 42°C (Note 4).
7. Wash slides with 2× SSC twice for 3 min at 42°C.
8. Take the staining jar containing slides out of the water bath or incubator and leave to cool for 5–15 min.
9. Proceed to counterstaining and mounting (for fluorochrome labeled probes, *Protocol 9.3*), or antibody detection (*Protocol 9.2*) or enzymatic detection (*Protocol 10.6*).

Notes

1. Molecular biology grade formamide will have a pH around 7 and be solid at −20°C; it degrades at room temperature. Very high quality formamide can be purchased frozen, but for the protocols here, freezing immediately on receipt is adequate if fresh from the supplier (Sigma catalogue number F7508 is suitable). Formerly, an ion exchange resin (e.g. Amberlite, Merck) was added to purify crude formamide grades not specified for molecular biology.
2. Adjust stringency as required. See *Table 7.2* and Chapter 7 on stringency.
3. This step controls the stringency of hybridization, must include at least one change of solutions, and must be accurate (±0.5°C).
4. Formamide needs to be washed away completely; inadequate washing is a source of background. Make sure that slides are covered totally with solution, otherwise dirt will accumulate at the top of the slide and run down to the preparation when changing solution or taking slides out.
5. Based on Schwarzacher *et al.* (1994).

Alternative washes without formamide for oligonucleotides or other probes

It can be desirable to avoid using formamide because of the hazards associated with the chemical, the costs and storage requirements. Formamide destabilizes the DNA double helix, and allows lower temperatures and higher salt concentrations to be used for the washes. As it is undesirable to raise the temperature to avoid loss of material, nor reduce the salt below 0.1× SSC to maintain some buffering, washes without formamide are usually carried out at a lower stringency than the hybridization. The wash solution used in Step 5 of *Protocol 9.1* is typically 0.1× SSC at a temperature of 45°C. For typical probes, this is about 76% stringency.

Because of the instability of hybrids involving short oligonucleotides, no formamide is usually used in the washes. Typically, slides are washed in 6× SSC (1.17 M Na^+) three times for 30 min at room temperature (Steps 3 and 4, *Protocol 9.1*), followed by the stringent wash in 6× SSC at 5°C below the respective duplex stability temperature (*T*m; Chapter 7) for 1–2 min (Step 5, *Protocol 9.1*).

9.2 Detection

Detection of hybridization sites is an important part of *in situ* hybridization allowing visualization of the probe–target hybrids formed (*Figure 9.1*). The detection steps required depend on the modified nucleotide used to label the probe (see *Figure 9.2*). If direct fluorophore-labeled dUTPs were used for labeling the probe, no immunohistochemistry for visualization is needed, and slides are counterstained and mounted immediately (*Protocol 9.3*). Probes labeled with biotin need to be detected with avidin, streptavidin, extra-avidin or antibodies to biotin, while those labeled with digoxigenin need anti-digoxigenin. Antibodies to fluorophores, in particular fluorescein, are also used frequently. The antibodies and avidins are conjugated to fluorophores, enzyme or metals giving three different ways for detection depending on the sensitivity and resolution required (see Section 4.1). Fluorescent detection systems are normally preferred for DNA *in situ* hybridization because of the precise localization of the hybridization signal and of the advantages of multi-target detection using the many different fluorophores available (see *Table 4.1*) and are described in *Protocol 9.2*. Colorimetric detection using antibodies conjugated to enzymes that catalyze the precipitation of color pigments from substrates (see *Figure 9.1*) are more laborious and are less spatially defined than fluorescence. However, they can be analyzed with transmitted light microscopy and can be more sensitive. They are most often used for RNA detection and are, together with the detection at the EM level, described in Chapter 10.

For fluorescent detection, most commonly FITC (blue excitation, yellow-green fluorescence), rhodamine, Texas Red or Cy3 (all green excitation, red fluorescence) are used. Other fluorophores used widely are coumarin (UV excitation, blue fluorescence) and Cy5 (red excitation, far red fluorescence), and the Alexa and Cy sets of fluorophores (see *Table 4.1*). If two or more probes are detected simultaneously, it is important to choose the fluorophores carefully, particularly

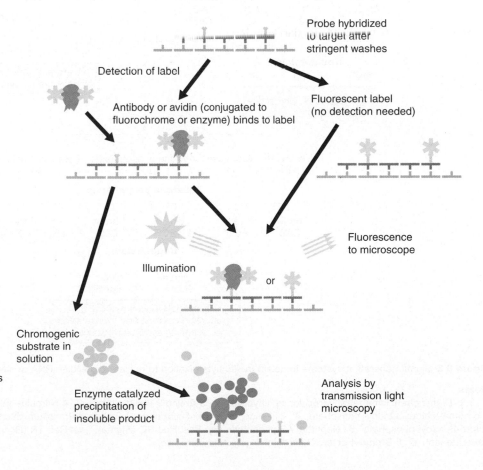

Figure 9.1 Detection of sites of hybridization of labeled probe using direct and indirect labels with fluorescent or enzymatic detection methods.

when several antibodies and detection reagents are mixed and used simultaneously (*Protocol 9.2*, Steps 3 and 4; see Schwarzacher and Heslop-Harrison 1994). Direct fluorophore-labeled probes can be combined with indirect labels, and are not affected by the detection steps for the indirect labels.

The antibodies and avidins used in the detection steps have very high affinity and high specificity for the label molecules. Hence they are easier to use and more robust than anti-protein antibodies and can be mixed without interference and loss of detection power in multi-target experiments. However, occasional batches from otherwise reliable suppliers do not work, and concentrations may alter by up to 10-fold between batches without notice from the supplier.

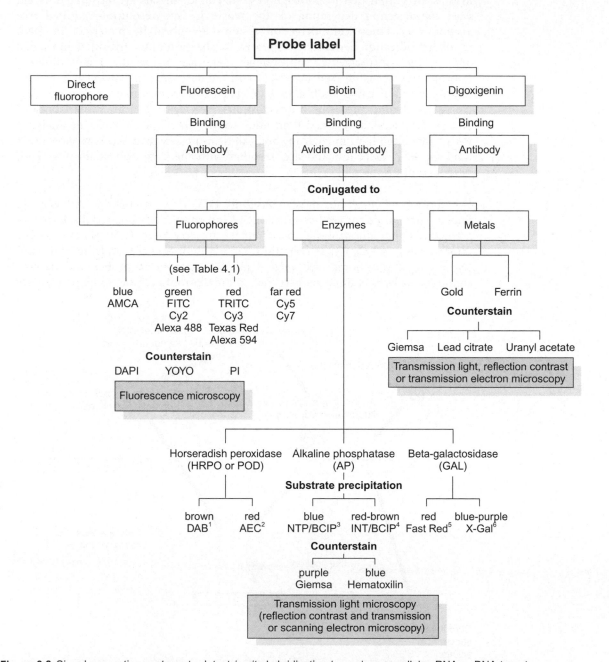

Figure 9.2 Signal generation systems to detect *in situ* hybridization to nuclear or cellular, RNA or DNA targets.

Notes
1. 3,3′-Diaminobenzidine tetrahydrochloride dihydrate. 2. 3-Amino 9-ethylcarbazol. 3. 4-Nitroblue tetrazolium chloride and 5-bromo-4-chloro-3-indoyl-phosphate. 4. 2-(4-Iodophenyl)-5-(4-nitrophenyl)-3-phenyltetrazolium chloride and 5-bromo-4-chloro-3-indoyl-phosphate. 5. Naphthol-AS-phosphate and Fast Red RC (Sigma), Fast Red TR (Roche) or Vector-Red (Vector Laboratories). 6. 5-Bromo-4-chloro-3-indolyl-β-D-galacto-pyranoside.

PROTOCOL 9.2 FLUORESCENT DETECTION OF HYBRIDIZATION SITES

Reagents

Detection buffer: 4× SSC (dilute from 20× SSC,
  Appendix 1) containing 0.2% (v/v) Tween 20.
BSA block: 5% (w/v) BSA (bovine serum albumin) in
  detection buffer (Note 1).
Detection solution: select as required (see *Figure 9.2*
  for explanation). Different detection reagents can
  be combined to detect different labels
  simultaneously.
  (i) For digoxigenin-labeled probes: anti-
       digoxigenin (FAB fragment) conjugated to

fluorescein or rhodamine (200 µg ml^{-1};
Roche), final concentration 2–4 µg ml^{-1} in BSA
block (Notes 2 and 3).
(ii) For biotin-labeled probes: avidin, streptavidin or
      extra-avidin conjugated to fluorescein, Texas
      Red, Cy3, Alexa fluorphores or others (e.g.
      Vector Laboratories, Sigma, Molecular Probes),
      final concentration 5–10 µg ml^{-1} in BSA block
      (Notes 2 and 3).

Material

Preparations labeled with biotin, digoxigenin or other
label to be detected following stringent washing

(*Protocol 9.1*) or *in situ* primer extension (*Protocol
8.5*).

Method

For the washing steps (1 and 5), slides are put into staining jars and covered
totally with solution. Agitation by gently moving the staining jars by hand
(once every minute) or continuously on a shaking platform is recommended.
To change solutions, pour off carefully and replace slowly with next solution,
or transfer slides to another staining jar with the next solution.

It is very important that slides do not dry out between steps.

Procedures using fluorophores should be carried out in subdued light to
avoid degrading the molecules.

1. Place slides in detection buffer and leave in buffer for 5 min.
2. Apply 200 µl BSA block to the marked specimen area of each slide and
    cover with a large plastic coverslip. Incubate at room temperature or
    37°C or 5–30 min in a humid chamber (Note 3).
3. Decide which detection reagents (see *Figure 9.2*) are required for which
    slides. Calculate and make up detection solutions: For example, for eight
    slides, each labelled with biotin and digoxigenin, 3 µl FITC-anti-
    digoxigenin (200 µg ml^{-1}, Roche), and 1 µl Cy3-Streptavidin
    (1 mg ml^{-1}, Sigma) in 300 µl BSA block will be suitable (see also Note
    4).
4. Put the coverslip to one side and drain BSA block away. Add 30–40 µl of
    detection solution. Replace the coverslip and incubate for 1 h at 37°C in
    a humid chamber.
5. Remove coverslips and wash in detection buffer for 20 min at 42°C for
    exchanging the solution three times.
6. Proceed to staining (*Protocol 9.3*).

Notes

1. Numerous qualities of BSA (50 from Sigma alone) are available; a mid-priced fraction V, often essentially globulin-free, specified for use in molecular biology and immunocytochemistry protocols is suitable, such as Sigma A7638, A3803 or B4287. BSA solutions must not be autoclaved as the protein coagulates.

2. Some researchers use anti-digoxin from Sigma. Reagents to detect biotin are numerous and are available conjugated to many fluorophores (see *Table 4.1*).

3. These are typically 1:50 to 1:500 dilutions of the stock solutions supplied by manufacturers. Concentrations must be optimized and vary with antibody batches; two-fold concentration changes are used during optimization.

4. This steps blocks non-specific sites that could bind detection reagents. Longer incubation at 37°C will result in better blocking, but is normally not necessary.

5. Based on Heslop-Harrison *et al.* (1991).

9.3 Visualization of DNA and mounting

DNA is usually stained with suitable fluorochromes that can be visualized separately or together with the *in situ* hybridization signal (counterstaining). Where multiple labels are used, or where background hybridization signal is high enough (often the case when using genomic DNA as a probe), counterstaining may not be required, and can reduce contrast of the signal or obscure important minor signals.

The preparation must be mounted in a suitable solution that prevents the fading of fluorescence when viewed (see also Chapter 11). Antifade solutions are essentially glycerol that on its own stabilizes fluorescence, but often other specific reagents are added. Many antifades are commercially available, e.g. Vectashield (Vector), Slow Fade (Molecular Probes), Fluorguard (Sigma) or Citifluor AF1 (Agar) or can easily be made (*Protocol 9.3*). High quality glycerol (ideally a quality specified for fluorescence applications) must be used since some grades absorb UV or auto-fluoresce. The nature and hydration level of the specimen, mountant refractive index and thickness, coverslip glass and thickness, immersion oil, and microscope lens characteristics, are all interdependent and must be optimized together to obtain signal near the theoretical resolution of the microscope.

PROTOCOL 9.3 FLUORESCENT STAINING OF DNA AND MOUNTING

Reagents and equipment

McIlvaine's buffer (pH 7.0): 82 ml of 200 mM Na_2HPO_4 and 18 ml of 100 mM citric acid.
$1\times$ PBS (Appendix 1).
Detection buffer: see *Protocol 9.2*.
Choose as required (Note 1):
  (i) DAPI (4′,6-diamidino-2-phenylindole, Sigma): prepare DAPI stock solution of 100 µg ml^{-1} in water. DAPI is a potential carcinogen. To avoid weighing out the powder, order small quantities and use the whole vial to make the stock solution. Aliquot and store at $-20°C$ (it is stable for years). Prepare a working solution of 2–4 µg ml^{-1} by dilution in McIlvaine's buffer, aliquot and store at $-20°C$ (Notes 2 and 3).
  (ii) PI (propidium iodide, Sigma): prepare PI stock solution of 100 µg ml^{-1} in water. PI is a potential carcinogen. To avoid weighing out the powder, order small quantities and use the whole vial to make the stock solution. Aliquot 50 or 100 µl in 1.5 ml microcentrifuge tubes and store at $-20°C$. Dilute with 1x PBS to 0.1– 5 µg ml^{-1} (initially, use a low concentration such as 0.3 µg ml^{-1}) prior to use. PI does not keep in diluted form (Note 2).

Antifade solution and mounting medium: these must be UV transparent and autofluorescence free.
  (i) Commercial mounting media: e.g. Vectashield (Vector Laboratories), Slow Fade (Molecular Probes), Fluorguard (Sigma) or Citifluor AF1 (Agar).
  (ii) To 60–90% (v/v) glycerol (specified for fluorescence) in $1\times$ SSC (or $1\times$ PBS or water) add one or more of the following:
    • 0.2–0.5% PPD (2,5-diphenyl-1,3,4-oxadiazol; Sigma).
    • 2–5% DABCO (1,4-diazabicyclo[2.2.2]octane; Sigma).
    • 4% *n*-propyl gallate (NPG).
  Store at 4°C; some solutions go brown with time but this does not effect their efficiency.
Glass coverslips: No. 0, 24 × 40 or 24 × 50 mm (Note 4).
Flat slide storage trays: the mountant remains fluid, so slides should be stored flat, in the dark in a cold room or refrigerator. Cardboard trays hold more slides than plastic and can be stacked better, but decay in the humid conditions.

Material

Preparations following stringent washes (direct fluorophore labels, *Protocol 9.1*), detection of hybridization sites (indirect labels, *Protocol 9.2*) or *in situ* primer extension (direct fluorophore labels, *Protocol 8.5*).

Methods

1. Remove slides from the staining jar and drain excess buffer.
2. Put 100 µl DAPI or PI solution on the marked area of the preparation on each slide. Cover with a plastic coverslip and incubate at room temperature for 10 min, avoiding bright light (Notes 2 and 3).
3. Remove coverslips and wash slides briefly in detection buffer.
4. Add two drops (total about 80 µl) of antifade solution to each preparation and cover with a large, thin coverslip avoiding bubbles. Allow to settle for a few minutes and then gently, but firmly, squeeze excess antifade from the slide between several sheets of filter paper.
5. Analyze preparations with an epifluorescence microscope (see Chapter 11).

Notes

1. Take care that the counterstains complement the hybridization detection fluorophores: do not use DAPI with blue fluorophores or PI with red fluorophores. It may also be worth inspecting slides without counterstaining since some probes (e.g. genomic DNA) label all chromosomes enough to

enable visualization. See *Table 4.1* for properties and further choices of DNA stains.
2. Avoid over-staining of chromosomes, which obscures weak hybridization. Concentration of stain, not time of staining, needs to be adjusted. Often, it is worth examining some slides

unstained. PI can be washed out by removing the coverslip and soaking for 5 min to overnight in 4× SSC/Tween; washing also reduces DAPI staining.

3. DAPI has two modes of binding to DNA: in the minor groove, which occurs at high stain concentrations; and, at lower concentrations, intercalation between AT-nucleotide tracts (Wilson *et al.* 1990). The latter binding gives the greater increase in fluorescence compared to the unbound molecule. Hence, increasing stain concentration has sometimes the unusual effect of reducing signal, and weak signal may mean too high a stain concentration.

4. No. 0 (0.1 mm) is thinner than the most widely used specification (No. 1, 0.13–0.16 mm) and is recommended for high resolution fluorescence microscopy.

Alternative protocol using antifade containing DNA counterstain

Some protocols use fluorescent DNA stains in solution in the mountant, and pre-prepared mixtures are available (e.g. Vysis, Sigma or Appligene). This removes Steps 2 and 3 from *Protocol 9.3*, and is worthwhile for routine applications, although stain in solution may reduce contrast, and staining is less easily varied. Add DAPI (e.g. 0.5 μg ml^{-1}) or PI (0.3 μg ml^{-1}) to the antifade mountant or buy a commercial mixture and mount as in Step 4.

9.4 Amplification of hybridization signal

Further rounds of staining with secondary antibody–fluorophore conjugates can amplify hybridization signal. Amplification (*Protocol 9.4*) can be carried out directly after detection of hybridization sites before staining, or after DAPI staining and examination under the microscope (remove coverslips and wash as for reprobing, *Protocol 8.6*). However, our experience is that background is amplified as much or more than the signal, and little signal is revealed that was not previously visible under optimum microscopy conditions, so we do not recommend routine use of amplification.

PROTOCOL 9.4 FLUORESCENT AMPLIFICATION OF HYBRIDIZATION SITES

Reagents and equipment

Detection buffer: 4× SSC (dilute from 20× SSC, Appendix 1) containing 0.2% (v/v) Tween 20.
Serum block: 5% (v/v) normal goat, rabbit or other serum depending on the amplification antibody used (see below) in detection buffer.
Amplification solution: select as required (Note 1)
  (i) For digoxigenin: e.g. 25 µg ml^{-1} FITC or TRITC anti-sheep IgG conjugate raised in rabbit (DAKO) in appropriate (sheep or goat) serum block.

  (ii) For FITC or other fluorophores: e.g. anti-fluorescein conjugated to FITC (DAKO) in appropriate serum block.
  (iii) For biotin: 5 µg ml^{-1} biotinylated anti-avidin (e.g. raised in goat, Vector Laboratories) in appropriate (e.g. goat) serum block.
BSA block and detection solution as in *Protocol 9.2*.

Materials

Preparations following detection of hybridization sites (direct labels, *Protocol 9.1*; indirect labels, *Protocol 9.2*) or counterstaining and mounting (*Protocol 9.3*).

Method

For the washing steps (3 and 7), slides are put into staining jars and covered totally with solution. Agitation by gently moving the staining jars by hand (once every minute) or continuously on a shaking platform is recommended. To change solutions, pour off carefully and replace slowly with next solution, or transfer slides to another staining jar with the next solution.

It is very important that slides do not dry out between steps.

Procedures using fluorophores should be carried out in subdued light to avoid degrading the molecules.

For slides that have been mounted and examined, start at Step 1. For slides directly from *Protocols 9.1*, or *9.2*, start at Step 4.

1. Blot immersion oil off surface of coverslip and carefully wipe around coverslip with solvent to remove any oil from the slide. Wipe underneath of slide.
2. Put slides in 37°C for 5–10 min to reduce the viscosity of the glycerol/antifade mountant. Remove coverslip by lifting slowly but steadily with the edge of a razor blade (Note 2).
3. Wash slides in detection buffer at room temperature for 5 min then twice for 15 min.
4. Add 200 µl of serum block to the marked area of each slide, and cover with a plastic coverslip. Incubate for 5–30 min at room temperature or 37°C (Note 3).
5. Remove the coverslip and drain serum block away. Add 30–40 µl of appropriate amplification solution. Replace the coverslip and incubate for 1 h at 37°C.
6. Remove coverslip and wash in detection buffer for 20 min at 42°C exchanging the solution three times.
7. If biotin-avidin is amplified continue with Step 8, otherwise proceed to *Protocol 9.3* to counterstain and mount preparation.
8. Apply 200 µl BSA block to the marked area of each slide and cover with a large plastic coverslip. Incubate at room temperature or 37°C for 5–30 min in a humid chamber (Note 3).

9. Put the coverslip to one side and drain BSA block away. Add 30–40 μl of detection solution. Replace the coverslip and incubate for 1 h at 37°C in a humid chamber.

10. Remove coverslips and wash in detection buffer at 42°C for 20 min exchanging the solution three times.

11. Proceed to *Protocol 9.3* to counterstain and mount preparation.

Notes

1. For the amplification of primary antibodies, a secondary antibody raised against IgG from the animal that was used to raise the primary antibody, conjugated to the same fluorophore as used in *Protocol 9.2*. For amplification of biotin detected with avidins, biotinylated anti-avidin is used that is then detected with another round of avidin. Different antibodies can be combined to amplify different labels simultaneously.

2. Do not dip slides with immersion oil into solutions to remove coverslip; an oil film will form on the surface of the solution and adhere to the preparation when taking it out. It is not possible to wipe off oil from coverslips completely without the likelihood of moving the coverslip and damaging the preparation.

3. This step blocks non-specific sites that could bind detection reagents. Longer incubation at 37°C will result in better blocking, but is normally not necessary.

9.5 Chromosome banding, silver staining, immunostaining and *in situ* hybridization

Many chromosome analysis protocols are compatible with *in situ* hybridization and may be used sequentially. For example, C-banding protocols may be used either before or after standard *in situ* hybridization protocols, and the preparations examined and photographed after each protocol is finished. Chromosome banding with conventional stains or with fluorochromes (Heng and Tsui 1993) are efficient methods to identify chromosomes, and used very extensively in conjunction with *in situ* localization of low-copy genes on chromosomes (see almost any issue of journals carrying reports of mammalian gene mapping and cytogenetics, such as *Cytogenetics and Cell Genetics*). Because plant chromosome banding is generally less efficient than that for animals, its use is less common, but protocols using banding before (Jiang and Gill 1993) and after (Leitch and Heslop-Harrison 1992) can be used. BrdU banding is also compatible with *in situ* hybridization (Viegas-Pequignot 1992), while mammalian and plant banding patterns can be mimicked by some repetitive DNA probes such as Alu for human chromosomes (Lichter *et al.* 1990 and in Boehringer Mannheim 1992) or simple sequence repeats for plants (Cuadrado and Schwarzacher 1999). Where rDNA activity is being examined, silver staining may be used to identify active loci, while *in situ* hybridization can show all sites of the rRNA genes (Neves *et al.* 1995).

CHAPTER 10

In situ hybridization to RNA and DNA in cells, tissue sections or whole mounts

This chapter describes protocols for hybridization to detect expressed messenger RNA (mRNA), viral or bacterial nucleic acids in cells, tissue and whole organisms. Sections and whole mounts of paraformaldehyde or glutaraldehyde fixed material are used most frequently (see Chapter 6, *Protocols 6.3–6.9*). The results answer questions about gene expression and infection in relation to cell and tissue organization. Preparation pretreatments (*Protocols 10.1* and *10.2*) are required to allow penetration of probes and to reduce background. Centrifuge preparations (Cytospins; *Protocol 6.10*) or simple pathological cell smears can provide suitable target material, but alcohol:acetic acid fixed spread preparations (from the protocols of Chapter 5) are normally not suitable as RNA is destroyed during the fixation.

The equipment required for *Protocols 10.1–10.7* is described for *in situ* hybridization to chromosome preparations (Section 8.1). Plastic forceps (tweezers) are convenient to transfer slides between solutions. For the detection of RNA, precautions against RNase activity are very important at all times until the target RNA is stabilized by forming duplexes with the probe. Wear latex or plastic disposable gloves and treat glassware and solutions to remove RNase (see Section 6.1 and *Protocol 6.1*). Plastic coverslips cut from bags are RNase-free, but glass coverslips should be washed in acetone for 15 min and baked (see *Protocol 6.1*). Methods for making hybridization mixture, hybridization and washes (*Protocols 10.3–10.5*) are similar to DNA hybridization to chromosomal targets (Chapter 8), but preparations require different pretreatments, do not require denaturation when RNA target sequences are detected, and use modified, RNase-free, buffers.

This chapter presents colorimetric detection methods for different probe labels; these methods can be applied to whole mount and spread preparations following DNA or RNA target *in situ* hybridization. For whole mount preparations, fluorescent detection is normally not used because of the substantial auto-fluorescence of many tissues and embedding media and because sub-nuclear resolution is not required; hence colorimetric detection becomes the choice (Warford and Lander 1994; *Protocol 10.6*). Color precipitates can be accumulated over several hours to days, enabling detection of small amounts of RNA or DNA and providing greater sensitivity than fluorescent detection. Colorimetric detection can also be combined easily with routine staining procedures for transmitted light microscopy used in most pathology laboratories.

10.1 Pretreatments

Protease digestion

Proteolytic digestion of preparations with proteases, proteinase K, or pre-digested pronase E is used to remove partially proteins that increase background and interfere with probe access (*Protocol 10.1*), and has a minor function in degrading nucleases. Protease treatment is particularly needed after cross-linking fixation and improves signal from probes longer than 100 bp, but is not required where oligonucleotide probes are used. Most protocols prefer proteinase K as it requires no pre-digestion, its activity is reproducible and it is nuclease-free. However, pronase may work better for some tissue as it has a broader specificity and it is recommended for plant material. The optimal concentration and lengths of treatment for both proteases depends on the tissue and preparation and needs to be adjusted empirically. Acid treatment, detergents (see alternative to *Protocol 10.1*) and pepsin (see *Protocol 8.1*) can also be used on sections, but are generally gentler and less effective for tissue fixed with cross-linking agents. After protease digestion, most preparations need to be re-fixed in paraformaldehyde (*Protocol 10.1*, Step 4) to stabilize the tissue and preserve nucleic acids.

PROTOCOL 10.1 PROTEASE TREATMENT

To prevent the contamination of preparations with nucleases, gloves should be worn throughout the procedures. Use RNase-free or DEPC-treated water and solutions for the detection of RNA (see *Protocol 6.1*).

Reagents

1× PBS: dilute from 10× PBS stock (Appendix 1).

4% Paraformaldehyde fixative: in the fume hood, add 4 g of paraformaldehyde (EM grade) to 80 ml water, heat to 60°C for about 10 min, clear the solution with a few drops of 1 M NaOH, let cool down, add 10 ml of 10× PBS (see Appendix 1) and adjust the final volume to 100 ml with water. Adjust to pH 7.2 with 5 N H_2SO_4. It is convenient to prepare the fixative during Steps 2 and 3 below (Note 1).

Protease solution as required (Note 2):

(i) Proteinase K, DNase and RNase-free (from *Tritirachium album*; e.g. Sigma or Merck): make up stock of 500 μg ml^{-1} (5–10 U ml^{-1}) in 50 mM Tris-HCl pH 7.6 (Appendix 1) and store in aliquots at −20°C. Dilute with 50 mM Tris-HCl pH 7.6 to 0.5–20 μg ml^{-1} (Note 3).

(ii) Pronase E, protease from *Streptomyces griseus*, Type XIV (DNase and RNase-free; e.g. Sigma): make up stock of 40 mg ml^{-1} in water (160 U ml^{-1}). If nuclease contamination is possible, incubate for 4 h at 37°C. Store in aliquots at −20°C. Dilute with 50 mM Tris-HCl pH 7.6 containing 5 mM EDTA (Appendix 1) to 125 μg ml^{-1} (0.5 U ml^{-1}).

If using pronase E: Glycine solution: Make stock of 10% (w/v) in water. Store at 4°C. Dilute to 0.2% in 1× PBS

Optional: ethanol series: 70% (v/v), 90% (v/v) and 96% (v/v) ethanol in water.

Material

Paraffin or cryostat sections, whole mounts or centrifuged material (from *Protocols 6.4, 6.7, 6.8* and *6.9*). On glass slides, mark relevant area of material with a diamond pen so it can be seen quickly on wet slides (see *Protocol 8.1*).

Method

For the washing steps, slides are put into staining jars and immersed in solution (typically 80–100 ml for up to eight slides). Agitation by gently moving the staining jars by hand (once every minute) or continuously on a shaking platform is recommended. To change solutions, pour off carefully and replace slowly with next solution, or transfer slides to another staining jar with the next solution. Do not let slides dry out between changes of solutions.

Carry out steps at room temperature if not otherwise stated.

1. Transfer slides to Tris-HCl pH 7.6 and leave for 5 min.
2. Shake off solution and add 200 μl protease solution, cover with a coverslip and incubate for 10–30 min at 37°C (proteinase K) or at room temperature (pronase E) (Note 4).
3. To stop the protease treatment, wash slides in water at 4°C for 2× 10 min after proteinase K treatment; or wash in glycine solution for 2 min at room temperature followed by 1× PBS for 5 min at room temperature after pronase E treatment.
4. Immerse slides in 4% paraformaldehyde fixative for 20 min at 4°C or 10 min at room temperature.
5. Rinse slides with three changes of 1× PBS.
6. Directly proceed to acetylation (*Protocol 10.2*) or dehydrate slides in an ethanol series and air-dry and proceed to hybridization (*Protocol 10.4*). Do not dehydrate whole mount preparations.

Notes

1. This solution should be made up fresh, but can be stored at 4°C for a few days.
2. We recommend proteinase K for mammalian tissue and pronase E for plant tissue.
3. Choice of a proteinase K concentration is influenced by the type of cellular preparation being investigated and the type of nucleic acid to be localized. The table below assists selection of a proteinase K concentration range. Extended ranges of time or concentration can be used if weak or negative results, or high background, are observed.
4. The duration of the treatment can vary with different tissue and should be determined empirically.
5. Protocol supplied by Tony Warford (Cambridge Antibody Technologies, Cambridge) with plant modifications from Leitch *et al.* (1994a).

Target	Preparation	Proteinase K range (μg ml^{-1})
DNA	Centrifuge preparations, cryo-sections, smears	0.5–5
	Formaldehyde fixed, paraffin or resin embedded	15–50
RNA	Centrifuge preparations, cryo-sections, smears	0.5–5
	Formaldehyde fixed, paraffin or resin embedded	5–20

Alternative protocol using detergents or acids to permeabilize material

Protease treatments have been reported to be unnecessary for some preparation types and cannot be used if both *in situ* hybridization and immunodetection of proteins are to be used simultaneously. Instead, mild acid hydrolysis and incubation in detergent solution have been used to enhance probe access to the target sequences. These probably cause limited depurination of nucleic acids and partial removal of highly cross-linked acidic nuclear proteins. Omit the protease steps of *Protocol 10.1*, but treat material with paraformaldehyde. Then incubate in 1× PBS containing 0.25–0.5% (v/v) Triton X-100 and/or 0.25–5% (v/v) Nonidet-P40 followed by 20% acetic acid or 0.1 M HCl (5–10 min) and wash in 1× PBS.

Alternative protocols using freeze thawing cycles to permeabilize mammalian whole mounts

For whole mount or thick sections of mammalian material, repeated freezing and thawing has been found to be very effective prior to *in situ* hybridization of DNA probes to chromosomes (I. Solovei and T. Cremer, personal communication). After incubation in detergent (see alternative protocol above), slides or coverslips with cells are incubated in 1× PBS containing 20% glycerol for 30–60 min. Remove and submerge into liquid nitrogen for a few seconds, let warm to room temperature and dip into glycerol; repeat freezing and thawing three to six times. Make sure slides do not dry out. Wash cells with PBS and treat with acid if needed.

Acetylation

Treatment of sections with acetic anhydride (*Protocol 10.2*) reduces the electrostatic binding of probe to sections by acetylation of positively charged amino groups such as basic proteins. Acetylation also prevents non-specific binding of the probe to poly-L-lysine-coated slides and removes endogenous biotin, which can both cause substantial background signal. Some protocols use this step routinely and we recommend it for plant tissue, but in some tissues and for the detection of DNA it may not be necessary. Acetic anhydride also serves to inactivate the proteases.

PROTOCOL 10.2 ACETYLATION

To prevent the contamination of preparations with RNase, gloves should be worn throughout the procedures. Use RNase-free or DEPC-treated water and solutions for the detection of RNA (see *Protocol 6.1*).

Reagents

Triethanolamine-HCl: make 2 M stock solution and adjust pH to 8 with HCl, dilute with water to 0.1 M, prior to use (Note 1).

Acetic anhydride: 100% liquid. Warning: This solution is highly combustible, volatile and corrosive and should be handled carefully.

1× PBS: dilute from 10× stock (Appendix 1).

Ethanol series: 70% (v/v), 90% (v/v) and 96% (v/v) ethanol in water.

Material

Paraffin or cryostat sections, whole mounts or centrifuged material (from *Protocols 6.4,*

6.7, 6.8 and *6.9*) after protease treatment (*Protocol 10.1*).

Method

The following steps are carried out at room temperature.

1. In the fume hood, place slides in a staining jar containing 0.1 M triethanolamine-HCl that is continuously mixed using a magnetic stirrer.
2. Add acetic anhydride to a final concentration of 0.25–5% (v/v), stirring very vigorously for 10 s.
3. Incubate slides for 10 min, stirring gently (Note 2).
4. Transfer slides to a new staining jar containing 1× PBS. Leave for 1–5 min.
5. Dehydrate slides in an ethanol series and air-dry. Do not dehydrate whole mount preparations.
6. Proceed to hybridization (*Protocol 10.4*).

Notes

1. Triethanolamine cannot be autoclaved.
2. Acetic anhydride is very unstable in water so use fresh solution if a second batch of slides needs to be treated
3. Protocol from Leitch *et al.* (1994a).

10.2 Hybridization mixture

Probes and labels

Three types of probes are used for RNA *in situ* hybridization: cloned DNA, DNA oligonucleotides or RNA (riboprobes), each with advantages in some experiments. Cloned double-stranded DNA is labeled with routine procedures (*Protocols 4.1–4.3*) and can produce long probes, giving greater sequence representation, but because of possible re-annealing of the two probe strands before binding to the target, can have poor hybridization efficiency. To overcome this problem, single-stranded probe DNA can be produced by PCR (Gyllenstein and Ehrlich 1988). Oligonucleotides, normally labeled by end labeling (*Protocol 4.4*), are single-stranded and have high efficiency of hybridization, but are only sensitive if used as cocktails of several labeled oligonucleotides complementary to different regions of the target molecule. RNA or riboprobes produced by *in vitro* transcription (*Protocol 4.5*) show a high efficiency of hybridization, but probes are less stable and require special cloning vectors. The use of sense-strand (control) and anti-sense probe RNA (complementary to the target, sense, mRNA) provides a valuable control (see Section 10.7).

Because of the problem of probe and detection reagent penetration into cross-linked tissue, the length of the probe after labeling is more critical than for hybridization to spread preparations. The optimum probe length depends on the tissue and its preparation: lengths of 50–200 bp are usually recommended, but up to 1 kb may work. Probe length is controlled by the synthesis step in nick translation and random primer labeling (*Protocols 4.1* and *4.2*), and appropriate primers can be chosen to determine the length of probe synthesized by PCR (*Protocol 4.3*). Alternatively probes can be cleaved by alkali hydrolysis (see *Protocol 4.5*) after labeling to yield probes of desired lengths.

As for DNA *in situ* hybridization to chromosomes, the choice of probe label depends upon the requirements for sensitivity and resolution of signal (Chapter 4). Radioactively labelled probes are widely considered more sensitive than non-radioactive methods, with disadvantages of extra stages involving photographic emulsions and development, safety, length of time for signal generation, and lower spatial resolution of signal. Although many commercially available probes for nonradioactive *in situ* hybridization are labeled with biotin, and detection with streptavidin is highly sensitive, biotin has the disadvantage of being widespread in tissues (Wood and Warnke 1981). Several methods have been described to block endogenous biotin, but these are not always efficient, so digoxigenin and fluorescein (as in the protocols here) are often preferred.

Hybridization mixture

Table 10.1 gives three examples of typical hybridization mixtures containing probes and reagents to control stringency, pH and ion concentration, and assist probe access. The concentrations suggested are typical and can be used in most situations, with optimization following the first experiment.

The functions and choice of many components are discussed in the stringency (Chapter 7) and DNA hybridization (Chapter 9) chapters. EDTA is present to chelate any Mg^{2+} ions, which activate nucleases. Nucleic acids (autoclaved DNA from *E. coli* or salmon sperm; tRNA from yeast) are used at high concentrations to block unspecific probe binding sites. Polyvinylpyrrolidone (PVP) and Ficoll (a non-ionic synthetic polymer of sucrose) increase the volume and the effective probe concentration, and may bind phenolic and other impurities. DNA probes are labeled by standard methods (see above). Agarose gel electrophoresis and comparison with the strength of standards (see Chapter 3) is used to estimate probe concentrations. Probes are made up in 50% formamide (*Protocol 10.3*) and, where required, denatured before adding to the hybridization buffer that often contains reagents (e.g. BSA, Ficoll and PVP) that should not be subjected to high temperatures.

Table 10.1 Hybridization mixtures used for cell, tissue or whole mount *in situ* hybridization. Three mixtures are shown, using labeled DNA probes to detect bacterial or viral DNA, labeled riboprobes to detect mRNA expression in plant tissue, and labeled oligonucleotides to detect mRNA (Note 1). The volumes are given for eight slides each using 30 µl hybridization mixture (20–100 µl or more may be used depending on tissue area). The hybridization mixture without probe (hybridization buffer) is usually used for pre-hybridization of material.

| Solution | Examples of hybridization mixtures | | | | | |
| | DNA probe to DNA | | Riboprobes to mRNA | | Oligonucleotides to mRNA | |
	Final concentration	Amount	Final concentration	Amount	Final concentration	Amount
100% Formamide (Note 2)	50%	75 µl (Note 3)	50%	75 µl (Note 3)	20%	75 µl
20× SSC (Appendix 1)	2×	25 µl				
10× Na-salts (Note 4)	390 mM Na$^+$		1× 315 mM Na$^+$	25 µl		
10× Modified PE buffer (Note 5) and 6 M NaCl					1× 618 mM Na$^+$	25 µl each
500 mM Tris-HCl, pH 6.8–7.5	1 mM	0.5 µl	10 mM	Note 6	50 mM	Note 6
50 mM EDTA	0.1 mM	0.5 µl	5 mM	Note 6	5 mM	Note 6
50% Dextran sulfate	10%	50 µl	10%	50 µl	10%	50 µl
100× Denhardt's solution (Note 7)			1×	2.5 µl		
Denatured salmon sperm DNA 1 µg µl⁻¹ tRNA 100 µg µl⁻¹	30 ng µl⁻¹	8 µl	1 µg µl⁻¹	2.5 µl		
Water (as required)		41 µl		15 µl		70 µl
SUBTOTAL, Hybridization buffer (Note 8)		**200 µl**		**200 µl**		**250 µl**
Hybridization buffer/slide		24 µl		24 µl		29.5 µl
Probe/slide (Note 9)	25–75 ng	6 µl	50–150 ng	6 µl	5–15 ng	0.5 µl
TOTAL VOLUME/SLIDE		**30 µl**		**30 µl**		**30 µl**
Hybridization temperature (Note 2)	42°C		50°C		37°C	

Notes

1. Different buffers are given for the detection of plant mRNA using riboprobes, bacterial or viral DNA targets in pathological tissue, and using oligonucleotides as probes although the buffers have similar components and can be exchanged to suit other material.

2. The formamide, SSC concentration and temperature determine the stringency of the hybridization (Chapter 7). A final concentration of 50% formamide and 2× SSC allows typical DNA:DNA hybrids of 80–85% homology to form at 42°C; the final concentration of 50% formamide and the given 1× Na-salts allows typical RNA:RNA hybrids of 75–80% homology to remain stable. The stability of hybrid molecules between target and oligonucleotide depends largely on the oligonucleotide length and their GC content (Chapter 7). Change the formamide concentration to

achieve hybridization at 10–20°C below the calculated T_m depending on estimated mismatch between probe and target. For a 20mer with 50% GC content, 20% formamide and the suggested salts will allow 5% (1 bp) mismatch.

3. The remaining formamide is added with the probe.

4. $10\times$ Na-salts: 3 M NaCl, 100 mM Na-phosphate buffer, 100 mM Tris-HCl pH 6.8, 50 mM EDTA.

5. $10\times$ PE-buffer: 500 mM Tris-HCl, pH 7.5, 45 mM $Na_4P_2O_7$, 2% (w/v) PVP (MW 4■,000), 2% (w/v) Ficoll (MW 400,000), 50 mM EDTA.

6. Tris-HCl and EDTA are added together with $10\times$ Na-salts or $10\times$ PE buffer, see Notes 4 and 5.

7. $100\times$ Denhardt's solution: 20% (w/v) Ficoll, 20% (w/v) PVP, 20% BSA in water.

8. This mixture is called hybridization buffer. If all slides require the same probe add 50 μl probe to make a total volume of 250 μl. Otherwise, distribute hybridization buffer accordingly into different tubes before adding individual probes.

9. Labeled DNA probes and riboprobes are prepared in 50% formamide at $5\times$ final concentration. Labeled oligonucleotides are in water at $60\times$ final concentration.

10. Protocols from Tony Warford and Leitch et al. (1994a).

PROTOCOL 10.3 HYBRIDIZATION MIXTURES FOR RNA AND DNA TARGETS IN CELLS

To prevent the contamination of preparations with nucleases, gloves should be worn throughout the procedures. Use sterile, RNase-free or DEPC-treated water and solutions for the detection of RNA (see *Protocol 6.1*).

Reagents

Select as required:

Formamide: molecular grade, store at −20°C in small aliquots (Note 1). Formamide is a potential carcinogen.

20× SSC (Appendix 1).

50 mM EDTA (Appendix 1).

Tris-HCl pH 7 (Appendix 1).

10× Na-salts: make up stock of 3 M NaCl, 100 mM Na-phosphate buffer (Appendix 1), 100 mM Tris-HCl, pH 6.8, 50 mM EDTA.

100× Denhardt's solution: 20% (w/v) Ficoll (MW 400,000), 20% (w/v) PVP (MW 40,000), 20% BSA in water; 0.22 μm filter sterilize and store at −20°C.

10× PE-buffer: 500 mM Tris-HCl, pH 7.5, 1% (w/v) $Na_4P_2O_7$, 2% (w/v) PVP (MW 40,000), 2% (w/v) Ficoll (MW 400,000), 50 mM EDTA in water; dissolve by heating to 65°C; store at room temperature.

50% (v/v) dextran sulfate in water; dissolve by heating to 65°C; 0.22 μm filter sterilize and store at −20°C (Note 2).

Sonicated or autoclaved DNA from salmon or herring sperm, or *E. coli* (1 μg μl^{-1}; treat DNA following *Protocol 3.2*, or purchase commercially). Fragments should be 100–300 bp long. Store in aliquots at −20°C.

Yeast tRNA: (e.g. Sigma Type X) 100 μg μl^{-1} in water.

Labeled and tested probe (Chapter 4) DNA or RNA: 50–100 ng μl^{-1}; make up to appropriate 5× final concentration in 50% formamide by adding formamide and water (e.g. 1 μl probe, 2 μl water and 3 μl formamide). Store at −20°C.

Oligonucleotides: 10–30 ng μl^{-1} (60× final concentration). Store at −20°C.

Method

This protocol can be followed in parallel to the pretreatment protocols.

1. Decide on probes and controls to be used (see Section 10.7) and record on *Form 10.1*.
2. Prepare appropriate hybridization buffer according to *Form 10.1*. Hybridization buffer without salmon sperm DNA can be stored at −20°C for further experiments.
3. DNA or riboprobes: for each slide, denature 6 μl probe–formamide mixture at 80°C for 2 min and cool on ice. Add 24 μl hybridization buffer to make the hybridization mixture and store on ice (Note 3). Oligonucleotides: for each slide, add 0.5 μl probe to 29.5 μl hybridization buffer and store on ice (Note 3).
4. Proceed to *Protocol 10.4* for hybridization.

Notes

1. Molecular biology grade formamide will have a pH around 7 and is solid at −20°C; it degrades at room temperature. Very high quality formamide can be purchased frozen, but good grade (e.g. Sigma catalogue number F7508), frozen immediately on receipt is adequate if fresh from the supplier.
2. The solution is very difficult, but possible, to force through the filter. To pipet the dextran sulfate

solution, cut off the end of the micropipet tip to give a larger orifice.

3. Typical hybridization mixture volume is 30 μl per slide, but 20–100 μl can be used depending on area of preparation on slide. Pool mixtures if more then one slide requires the same probe.

4. Protocol from Leitch *et al.* (1994a), including information from Tony Warford.

FORM 10.1: *Whole Mount* In Situ *Hybridization Data Sheet*

RUN CODE: Name: Date: .   .   .   20

Cross-reference to notes or photographs of slides or materials:

Component	Typical value (final concentration) volume	Hybridization buffer			
		A	B	C	
(delete as required)		DNA to DNA	RNA to RNA	Oligo-nucleotides	
100% Formamide	(20–50%) 75 µl				
20× SSC	(2×) 25 µl				
10× Na-salts	(1×) 25 µl				
10× PE	(1x) 25 µl				
500 mM Tris-HCl pH 7.2	(1 mM) 0.5 µl				
50 mM EDTA	(0.1 mM) 0.5 µl				
6 M NaCl	(600 mM) 25 µl				
50% Dextran sulfate	(10%) 50 µl				
100× Denhardt's solution	(1×) 2.5 µl				
Salmon sperm DNA 1 µg µl^{-1}	(1 µg) 8 µl				
tRNA 100 µg µl^{-1}	(0.15%) 2.5 µl				
Water (as required)					
TOTAL	200 (A, B) or 250 (C) µl				

		Slide code or details									
Component	Typical value	1	2	3	4	5	6	7	8		
Hybridization buffer	24 (A, B) or 29.5 (C) µl										
Probe*	6 or 0.5 µl										
Control probe*	6 or 0.5 µl										
No probe											
Pretreatment											
Total volume											

*Insert details and concentration

Pre-hybridization:

Denaturation times and temperatures for the preparations:

Hybridization time and temperature: Washing conditions:

10.3 Hybridization

Pre-hybridization

For *in situ* hybridization to whole mount preparations, material is pre-hybridized by incubation in the hybridization buffer (not containing the probe) for 1–2 h at the hybridization temperature (*Protocol 10.4*, Step 1), ensuring penetration of the tissue and blocking of nonspecific binding sites. For sections, this step is not necessary and sometimes can lead to decreased signal as the small amounts of probe are diluted. Some protocols prefer to dehydrate sections prior to applying the hybridization mixture to ensure accurate concentrations (see *Protocol 10.1*, Step 6 and *Protocol 10.2*, Step 5). Sometimes this step destroys the material (whole mounts) or embedding media (acrylic resins) or generates background and should be omitted.

Hybridization

Double-stranded DNA target sequences need to be denatured prior to hybridization. This is best achieved by applying the hybridization mixture to the material and heating it together to 85–95°C as described for spread chromosomal DNA targets (*Protocol 8.4*) although slightly higher temperatures are used after cross-linking fixation. RNA targets are single-stranded and do not need to be denatured.

PROTOCOL 10.4 HYBRIDIZATION TO CELLS, SECTIONS OR WHOLE MOUNTS

Reagents

Coverslips: No. 1 22 × 22 mm or suitable size to cover the preparation, wash in acetone for 15 min and bake (see *Protocol 6.1*) or use plastic coverslips cut

from new bags used for autoclavable waste (see Section 8.1)
Optional: appropriate hybridization buffer used to make up hybridization mixture (see *Protocol 10.3*).

Materials

Fixed and pretreated cell, tissue section or whole mount preparations from *Protocol 10.1* or *10.2*.

Single-stranded or denatured hybridization mixture from *Protocol 10.3*.

Method

If pre hybridization is required, in particular for whole mount preparations, start at Step 1; otherwise start at Step 2.

1. Apply 200 µl of hybridization buffer to each marked area on preparations. Cover with coverslip and incubate at the hybridization temperature (see *Table 10.1*) for 30 min to 2 h. Remove coverslip and drain excess fluid, avoid drying out of the preparation.
2. Add hybridization mixture to each marked area on preparations. If slides were incubated in buffer, shake off excess fluid first; avoid drying out of the preparation. Cover with a suitable size of coverslip. Make sure no bubbles are trapped underneath the coverslip (Note 1).
3. If denaturation of target sequences is required, heat slides on a heated block at 85–95°C, leave them for 5–10 min and then cool them down to the hybridization temperature (slow cooling is usually used) (Note 2).
     Thermal cyclers have the most accurate temperature control, but a heated plate with good temperature control can be used. Alternatively, a water bath at 90°C with a lid (essential to maintain a high temperature) can be used. Slides are rested on glass rods lying on moist paper in a closed heat-resistant chamber floating in the water bath, with a digital thermometer probe inside the chamber next to the slides. The temperature of the chamber is equilibrated and the temperature within the chamber accurately controlled by opening and closing the lid of the water bath during denaturation.
4. Incubate slides at the hybridization temperature (see *Table 10.1*) in a humid chamber, in the thermal cycler or on the heated block, typically for 16–20 h (Note 4). Check slides do not dry out or accumulate condensation.
5. Proceed to stringent washing (*Protocol 10.5*).

Notes

1. For hybridization times up to 20 h, sealing of coverslips is normally not necessary. However, for hybridization times of several days or when using very small amounts under a glass coverslip, seal with rubber cement to avoid drying out.
2. The range of time and temperature has to be adjusted for different species and material: 90–95°C for 5 min provides conditions that denature most DNA. During denaturation, it is important that no condensation drips on to the slides or the

hybridization mixture will be diluted and stringency decreased.
3. Hybridization time depends on the amount and length of probe and amount of target. For highly abundant sequences, 4–6 h will be adequate, but it can be advantageous to leave rare target sequences to hybridize for 72 h or more.
4. The protocol described here is based on Heslop-Harrison *et al.* (1991), Leitch *et al.* (1994a) and information received from T. Warford.

10.4 Post-hybridization washes

The washes after hybridization remove unbound or weakly bound probe, and the hybridization mixture. Typically, the washes are carried out at lower or the same stringency as hybridization, but are longer than for DNA hybridization to chromosomal spreads because the washes should wash away, where possible, probe that is bound nonspecifically to proteins.

10.5

PROTOCOL 10.5 POST-HYBRIDIZATION WASHES FOR DNA AND RIBOPROBES

PROTOCOL

Reagents

20× SSC stock (Appendix 1).

Optional formamide: good but not the highest grade. Immediately after purchase, store at −20°C in aliquots sufficient for the washes in one experiment, typically 40 ml (Note 1). Formamide is a potential carcinogen.

Stringent wash solution: the same or lower stringent conditions than hybridization are recommended (Note 2). Use 0.1× SSC, optionally containing 2 mM MgCl$_2$ and 0.1% Triton X-100 or Tween 20. Alternatively include formamide to a final concentration of 20% in above solution (i.e. 40 ml

100% formamide and 1 ml 20× SSC made up to 200 ml).

1× and 2× SSC: diluted from 20× SSC.

1× PBS or 1× TBS (Appendix 1).

Optional 10× NTE buffer (Appendix 1).

Optional RNase solution: DNase-free ribonuclease A (e.g. Sigma R4642, solution in 10 mM Tris-HCl, pH 8 and 50% glycerol, 70 units per mg protein). Make up stock of 10 mg ml^{-1} in 10 mM Tris-HCl, pH 8 (Appendix 1). Store at −20°C in aliquots. Dilute to 100 µg ml^{-1} in 1× NTE buffer prior to use (Note 3).

Method

For the washing steps (4–8), slides are put into staining jars and covered totally with solution. Use a gentle shaking water bath, rotating platform or hand-agitate the solution. To change solutions, pour liquid off carefully and replace slowly with next solution, avoiding turbulence, or transfer slides to another staining jar with the next solution. Do not let slides dry out between changes of solutions.

1. Prepare stringent wash solution and heat to 42–50°C in a water bath. Typically, for eight slides in a 100 ml staining jar, you will need 300 ml.
2. Take slides from hybridization chamber, avoiding condensation dropping on to the slides. Examine slides and note anything unusual: Have any slides dried out? Has any water dropped on the slides? Are there any bubbles?
3. Float off coverslips by incubating slides in a staining jar with 2× SSC at 30–40°C. The plastic coverslips will fall away in the solution (this might take up to 30 min) and are removed with a pair of forceps. Any surface film or dust must be wiped from the liquid in the jar with a tissue before pouring away solution. Otherwise it will stick strongly to preparations.
4. Incubate preparations in stringent wash solution twice for 20–90 min at the hybridization temperature (usually at 42 or 50°C, see *Form 10.1*) shaking gently. Measure the temperature in the solution in the staining jar, and remove from water bath or heat to maintain temperature to within 0.5°C (Notes 2 and 4).
5. For removing unhybridized RNA, wash preparations in NTE buffer twice for 5 min at 37°C; apply 200 µl of RNase solution to each marked area, cover with a coverslip and incubate at 37°C for 30 min in a humid chamber. Wash in NTE buffer twice for 5 min at room temperature (Notes 5 and 6).
6. Repeat Step 4 incubating preparations in stringent wash solution for

20–60 min at the hybridization temperature, shaking gently. Measure the temperature in the solution in the staining jar, and remove from water bath or heat to maintain temperature to within 0.5°C (Notes 2 and 4).
7. Wash slides in 1× SSC for 2× 10 min at room temperature (Note 7).
8. Wash slides in 1× PBS or 1× TBS for 5 min.
9. Proceed to antibody detection (*Protocol 10.6*). For choices see *Figure 9.1*.

Notes

1. Molecular biology grade formamide will have a pH around 7 and be solid at −20°C; it degrades at room temperature. For the protocols here, freezing immediately on receipt is adequate if fresh from the supplier (Sigma catalogue number F7508 is suitable).

2. For most experiments using the hybridization mixtures and temperatures of *Table 10.1*, post-hybridization washing in 0.1× SSC at 42°C for DNA detection and 50°C for RNA detection will be sufficient. However, if high background or non-specific signals are experienced, increase the stringency by including formamide in the wash. Raising temperature much above 60°C is not recommended. For calculation of stringency see Chapter 7.

3. The RNase must be DNase-free. If RNase is not purchased DNase-free, inactivate DNase in the RNase by placing the stock solution in boiling water for 15 min.

4. This step reduces background and unspecific binding. If formamide is included and higher temperatures are chosen it also controls the stringency of hybridization and must be accurate

(±0.5°C). A total of three incubations are given here, two before (Step 4) and one after (Step 6) the optional RNase treatment totaling some 2.5 h. Some experiments will not need such extensive washes, and some protocols only use stringent washing after the RNase treatment. We recommend at least two changes with the minimum time to achieve clean slides; prolonged and stringent washes damage the preparations.

5. This step is omitted for the detection of DNA.

6. Be careful with RNase cross-contamination: the enzyme is extremely stable and difficult to inactivate. Do not use equipment or glassware that is used for probe preparation, pretreatment or hybridization.

7. Formamide needs to be washed away completely; inadequate washing is a source of background. Make sure that slides are covered totally with solution, otherwise dirt will accumulate at the top of the slide and run down to the preparation when changing solution or taking slides out.

8. Protocol based on Leitch *et al.* (1994a) and information from Tony Warford

Alternative protocol for post-hybridization washing of oligonucleotide probes

Oligonucleotides need gentle washing steps to ensure that the short probes remain hybridized to the target. Follow *Protocol 10.5*, but use 4× SSC, 6× SSC or TBS buffer (Appendix 1; optionally containing 0.1% Triton X-100 or Tween 20) for the stringent washes (5–20 min). RNase treatment (Step 4) before or between the stringent washes is recommended when detecting RNA targets.

10.5 Colorimetric detection of hybridization sites

For detection of RNA and DNA hybridization sites in cells, tissue sections and whole mounts, chromogenic, enzyme-mediated reporter systems that generate insoluble colorimetric precipitates are preferred. These methods can also be used instead of fluorescent detection for DNA *in situ* hybridization to chromosome spreads. Antibodies (or avidin) conjugated to the enzyme are allowed to bind to the hybridized, labeled probe, and then incubated with a suitable chromogenic substrate for the enzyme (*Figures 9.1* and *9.2*). Most commonly, alkaline phosphatase (AP) conjugated to (strept)avidin, anti-digoxigenin or anti-fluorescein is used with the NBT/BCIP detection system (*Protocol 10.6*). Horseradish peroxidase (HRPO or POD), β-galactosidase and glucose oxidase are alternative, widely available, enzymes with suitable chromogenic substrates. Depending on the abundance of the target to be detected the reaction is left for a few minutes to hours and even days. Colorimetric detection can be extremely sensitive, allowing visualization of single molecules of nucleic acids, but is limited by the accumulation of non-specific background, and results are spatially more diffuse than fluorescent detection (see Chapter 9).

NBT and BCIP are used under alkaline conditions as chromogenic substrates for AP. A reaction initiated by cleavage of the phosphate group from BCIP by AP yields a blue color and reduces NBT to yield an insoluble purple precipitate. Alkaline phosphatase can also be detected with Fast Red or other colors (see *Table 9.1* and catalogues of suppliers of immunocytochemistry regents, e.g. DAKO, Vector Laboratories, Sigma, Roche). Horseradish peroxidase cleaves the substrate DAB (3,3′-diaminobenzidine tetrahydrochloride dihydrate) to yield a brown precipitate that is insoluble in alcohol, xylene and water.

Some tissues contain endogenous enzyme activity that can produce large amounts of unspecific background. Endogenous peroxidase can be blocked with periodate and borohydride (Heyderman, 1979), sodium nitroferricyanide (Straus, 1971) or phenyl hydrazine (Straus, 1972). Endogenous APs are found in many tissues and can be blocked by adding levamisole to the color substrate (see *Protocol 10.6*; Ponder and Wilkinson 1981). These methods are not completely effective and it is often better to change the signal-generating system if high levels of signal are seen in controls.

Double and multi-target hybridization is not often performed with colorimetric detection. Different haptens and color precipitates are available (see *Figure 9.1*) and are sold as sets (e.g. 'multicolor detection set' from Roche giving green, blue and red colors), but most have been developed for Southern or Western blotting, immunocytochemistry and ELISA technology rather than *in situ* hybridization. Because of the high sensitivity of AP and availability of several chromogenic substrates, multi-target hybridization often uses different haptens for probe labeling, and detects them sequentially with different AP conjugates. Between detection steps, the alkaline phosphatase activity is destroyed (see e.g. Fobert *et al.* 1996).

In situ hybridization with nucleic acid probes can be combined with immunodetection protocols for proteins, although proteinase pretreatments cannot be used (Jackson *et al.* 1992; Strehl and Ambros 1993; Vass *et al.* 1989).

Reagents

Detection buffer: 4× SSC (dilute from 20× SSC, Appendix 1) containing 0.2% (v/v) Tween 20, or TBS containing 0.1% (v/v) Triton X-100.

Blocking solution: 5% (w/v) BSA in detection buffer (Notes 1 and 2).

Detection solution: select as required and prepare a final concentration of 0.3–1.5 U ml^{-1} in blocking solution immediately before use (Note 3).

(i) For digoxigenin-labeled probes: anti-digoxigenin (FAB fragment) conjugated to AP (Roche).

(ii) For fluorescein-labeled probes: anti-fluorescein (FAB fragment) conjugated to AP (e.g. DAKO, Roche, Life Technologies, Novocastra).

(iii) For biotin-labeled probes: avidin, streptavidin or extra-avidin conjugated to AP (e.g. Vector Laboratories, Sigma, Roche, Amersham).

AP-substrate buffer: 100 mM Tris-HCl, pH 9.0, 50 mM MgCl$_2$, 100 mM NaCl.

AP-substrate solution: ready-made stock solutions of NBT (75 mg ml^{-1} in 70% dimethylformamide) and BCIP (50 mg ml^{-1} in 100% dimethylformamide) can be purchased separately or already mixed (e.g. Roche, Life Technologies). Alternatively make your own stocks and freeze in appropriate aliquots. For use dilute 90 µl NBT and 70 µl BCIP (or as recommended by the supplier) in 10–45 ml of AP-substrate buffer (Note 4).

Material

Preparations after post-hybridization washes (*Protocol 9.1* or *10.5*).

Method

For the washing steps (1 and 4), slides are put into staining jars and covered totally with solution. Agitation by gently moving the staining jars by hand (once every minute) or continuously on a shaking platform is recommended. To change solutions, pour off carefully and replace slowly with next solution, or transfer slides to another staining jar with the next solution.

It is very important that slides do not dry out between steps.

1. Place slides in detection buffer and leave in buffer for 5 min.
2. Apply 200 µl blocking solution to the marked area of each slide and cover with a large plastic coverslip. Incubate at room temperature or 37°C for 5–30 min in a humid chamber (Note 5).
3. Put the coverslip to one side and drain block solution away. Add 30–40 µl of detection solution. Replace the coverslip and incubate for 30–90 min at 37°C in a humid chamber (Note 6).
4. Remove coverslips and wash in detection buffer at 42°C for 20 min exchanging the solution three times (Note 6).
5. Incubate slides in AP-substrate buffer for 5–10 min.
6. Apply 200 µl AP-substrate solution to the marked area of the slides and incubate in the dark at room temperature for 15–60 min for highly abundant targets. Low abundant targets will need incubation for several hours to days; make sure that slides do not dry out by putting them in small trays and sealing them. Do not cover preparations with a coverslip (Note 7).
7. Wash slides in running tap water for 5 min and optionally pass through a graded ethanol series 70%, 95%, 100%, 95%, 70% and water, 1–5 s each (Note 8).
8. Proceed to staining (*Protocol 10.7*).

Notes

1. Numerous qualities of BSA (50 from Sigma alone) are available; a mid-priced fraction V, often essentially globulin-free, specified for use in molecular biology and immunocytochemistry protocols is suitable, such as Sigma A7638, A3803 or B4287. BSA solutions must not be autoclaved as the protein coagulates.

2. Alternatively or in addition to BSA, some protocols use 2–20% (v/v) normal goat or rabbit serum, or 0.5–5% (v/v) commercially available blocking reagents (e.g. from Roche or Amersham).

3. These are typically 1:500 to 1:3000 dilutions of the stock solutions supplied by manufacturers. Concentrations must be optimized and vary with antibody batches; two-fold concentration changes are used during optimization.

4. 1–5 mM levamisole (Sigma) can be added to stop endogenous phosphatase activity.

5. This step blocks non-specific sites that could bind detection reagents. Longer incubation at 37°C will result in more complete blocking, but is normally not necessary.

6. Some protocols perform antibody incubation and washing at room temperature.

7. The AP enzyme will continue to produce an insoluble blue-black precipitate over a period of several days. In practice, non-specific precipitation will become a problem if incubation is extended over 36 h.

8. The precipitate is alcohol and xylene soluble; passing quickly through an ethanol series washes off background.

9. Protocol provided by Tony Warford (Cambridge Antibody Technologies, Cambridge) and modified after Fobert *et al.* (1996).

Alternative protocol detecting AP-conjugated antibodies with Fast Red

Instead of the NBT/BCIP color substrate, several companies sell Naphthol-phosphate/Fast-Red systems, e.g. Fast Red RC (Sigma), Fast Red TR (Roche) or Vector-Red (Vector Laboratories). Make up solutions as directed by the manufacturer and incubate preparations (Step 6) for several hours to days changing the solution once or twice per day. Rinse slides in water.

Alternative protocol detecting hybridization sites with horseradish peroxidase (HRPO or POD)

Follow *Protocol 10.6*, but use horseradish peroxidase-conjugated antibodies at 0.5–2 U ml^{-1} in detection buffer (Step 3). DAB, a brown soluble reagent, forms an insoluble, dark precipitate with the action of peroxidases. Because DAB is very carcinogenic, pre-weighed vials or tablets are best to use: Sigma sells tablets with buffer and H$_2$O$_2$ components included. To make the solution for detection of enzyme sites by precipitation of the chromogenic reagent, add 5 mg DAB in 0.5 ml water (store frozen in aliquots) to 9.5 ml of 50 mM Tris-HCl pH 7.4 and incubate slides with 200 μl each in the dark for 20 min at 4°C. Drain slides and add a further 200 μl DAB solution containing 0.1–0.5 μl 30% H$_2$O$_2$ and incubate for 20 min at 4°C. Stop the reaction with excess water. DAB precipitates can be enhanced with silver staining.

10.6 Staining and mounting

Many staining protocols can be used to identify and contrast cellular components against the enzyme-catalyzed precipitate (e.g. toluidine blue, hematoxilin, eosin, methyl green, neutral red, safranin and Giemsa; see *Figure 9.2*). If there is enough contrast, material can be mounted unstained. Detailed procedures are described in histological protocol books (e.g. Gurr 1971 or Pearse 1972, although the 2nd edition is more protocol-oriented than the 3rd edition) and only a selection is given in *Protocol 10.7*. In general, staining should be weak and contrast with the *in situ* hybridization signal, although many chromogenic precipitates tend to be a shade of gray-brown. When choosing staining protocols and mountants, consider the permanence needed and any cross-reaction with the color precipitate: xylene-based mountants can cause crystal formation of NBT/BCIP.

PROTOCOL 10.7 COUNTERSTAINING AND MOUNTING FOR TRANSMITTED LIGHT MICROSCOPY

Reagents

Staining solution (use cytological or histological stain normally used to examine the tissue in question), for example:
  (i) Giemsa (to stain DNA): 4% (v/v) in Sörenson's buffer (30 mM KH_2PO_4 and 30 mM Na_2HPO_4. Make immediately before use. If precipitate develops on the surface, remove with filter paper.
  (ii) Calcofluor White (to stain cellulose in plant cell walls): e.g. Sigma Fluorescent Brightener 28. Make a 0.1% (v/v) solution in water, filter through 0.2 μm filter before use.
  (iii) Hematoxilin (to stain nuclear proteins in mammalian tissue): e.g. Mayer's hematoxilin.
Mounting media: aqueous glycerol-based media, Euparal, DePeX mountant, or xylene (Note 1).
Coverslips: No. 0 24 × 40 or suitable size to cover the preparation.

Material

Preparations after colorimetric detection (*Protocol 10.6*).

Method

1. Incubate slides in staining solution for a few seconds (hematoxilin) or 5–10 min (Giemsa and Calcofluor White).
2. Wash in running water for 2–5 min.
3. Add two or three drops water-based mountant or glycerine; alternatively, dehydrate through 70%, 90% and 100% alcohol before mounting with Euparal, or dehydrate and pass through xylene before mounting in a drop of xylene or other mountant (e.g. DePeX) (Note 1).
4. Cover with a coverslip and remove excess mountant with a filter paper, pressing gently but firmly.
5. View under the light microscope (see Chapter 11).

Note

1. For NBT/BCIP detection a water-based mountant should be used. Organic solvents may dissolve or crystallize some color precipitates. Xylene is a satisfactory mountant for viewing of DAB detection and Giemsa-stained material. After observation, store slide in the dark; the xylene will evaporate and coverslip fall off, but new xylene can be added.

10.7 Controls for RNA experiments

To assess the hybridization signal and to distinguish it from background, controls should be always included when detecting RNA and DNA in tissue sections and whole mounts. DNA:DNA *in situ* hybridization to chromosomes does not normally need such controls as the target is well defined, and signal, as double dots originating from each chromatid, is expected at the same position on each relevant chromosome, and in each cell. However, mRNA, viruses and bacteria do not have specific enough distributions to use 'internal standards' as controls, and many labels or enzymes are found endogenously in tissue (e.g. biotin, phosphatase) and can mimic positive results. Consecutive sections of the same tissue are ideal to allow comparison of control and test hybridization, but preparations on single slides can be divided into different regions or different preparations can be used.

Controls (see box below) include positive and negative samples, as well as technical controls to detect false positive and negative results: presence of endogenous biotin or enzymes and unspecific binding of the antibodies and detection reagents needs to be checked

---

Controls for in situ *hybridization to tissue sections and whole mounts*

POSITIVE CONTROLS (to check that the hybridization and detection have worked)

1. Tissue or cell line known **to contain** mRNA, or bacterial or viral nucleic acid of interest.
2. Labeled RNA, DNA or oligonucleotide probe complementary to abundant target e.g. 'housekeeping' genes such as α-tubulin.

NEGATIVE CONTROLS (to check the extent and distribution of non-specific hybridization and background caused by, for example, non-specific probe binding sites, endogenous biotin or endogenous enzymes).

1. For mRNA targets, carry out hybridization using a probe with a sequence identical to, not complementary to, the target sequence. This is the single most important control, and the result is often illustrated in publications since comparison with the complementary probe clearly shows the specificity of the

signal distribution obtained. Generation of single-stranded sense and anti-sense riboprobes is described in *Protocol 4.5*.

2. Tissue or cell line known **not to contain** mRNA, or bacterial or viral nucleic acid targets homologous to the probe.
3. Hybridization with a probe known not to be present in the tissue or cells (e.g. plasmid, viral or bacterial sequence) to show specificity of signal obtained.
4. Digestion of RNA with RNase prior to *in situ* hybridization to show no signal is obtained.
5. Hybridization without probe to check specificity of detection.
6. Hybridization with or without probe in combination with detection steps without antibody to show the enzyme in the antibody (or avidin) is not present endogenously.

10.8 *In situ* hybridization in the electron microscope

In situ hybridization for transmission electron microscopy (TEM) (e.g. Binder *et al.* 1986; Leitch *et al.* 1990; Wachtler *et al.* 1992) or scanning electron microscopy (e.g. Herrmann *et al.* 1996; Lehfer *et al.* 1991) is a powerful and high resolution method to localize DNA or sometimes RNA sequences at the sub-micron level. However, the technical difficulties are considerable (particularly with loss of sections due to the multiple washes and high temperatures for denaturation and washing, and deposits of dirt). Other methods should be used where lower resolution is adequate; sometimes, hypotheses made using EM *in situ* hybridization can be confirmed by analysis of larger numbers of samples by light microscopy. The best morphological preservation for TEM is obtained from extended fixation in cross-linking agents, which is not helpful for *in situ* hybridization, requiring retention of intact and accessible nucleic acid targets.

Generally, for TEM, material is fixed in paraformaldehyde or glutaraldehyde and embedded in hydrophobic acrylic resins before ultra-thin sectioning (*Protocol 6.6*). Consecutive sections are collected on gold or nickel grids and can be subjected to the protocols described in this chapter. Grids are treated and washed by submering, or floating section side down, in droplets of solution, as carried out for staining. For long incubations, use small humid chambers in Petri dishes and seal or cover the grids and solution with a piece of Parafilm or autoclavable bag to avoid drying out. Typically, hybridization sites are detected with the peroxidase–DAB precipitation method (see alternatives to *Protocol 10.6*), or using methods parallel to those in *Protocol 10.6* with antibodies or avidin conjugated to colloidal gold (available in different sizes such as 5, 10, or 20 nm), followed by no or conventional EM staining methods. If sections are not supported on a film but lie across the grid bars, double-target hybridization can be used with labeled probes being hybridized separately to each side of the section, followed by detection with different sizes of colloidal gold on each side (McFadden *et al.* 1990).

It is also possible to analyze *in situ* hybridization preparations in the light microscope before transfer to the TEM for high-resolution analysis. Protocols are given by Schwarzacher and Heslop-Harrison (1991). In the future, it is probable that alternative labeling systems, including heavy metals, may be useful for electron microscopy.

CHAPTER 11

Microscopy, analysis of signal and imaging

11.1 Analysis of *in situ* hybridization signal

The analysis required depends on the experiment. Normally, photographic or digital images are obtained from a microscope, and increasingly computer-assisted digital image analysis is used to assist interpretation and to prepare results for publication. After some 50 years of little development, nearly every aspect of the light microscope and light microscopy has improved during the 1990s, and we can expect this rapid development to continue well beyond 2000.

Images from light and epifluorescence microscopes are required for analysis of the great majority of *in situ* hybridization experiments, and key points are discussed in Sections 11.2 (microscope and lenses) and 11.3 (filters). In this chapter, we focus on a brief overview of methods using transmitted light and epifluorescence microscopes; other books (e.g. Bradbury and Bracegirdle 1998; Herman 1997) or manufacturers' manuals must be used for alternatives. Advanced derivatives of epifluorescence microscopes include confocal microscopes, which are used for several reasons: the convenience of obtaining digital images, the possibility to remove blur from the image, and sometimes, but most importantly, for three-dimensional analysis of the distribution of targets in thick sections or whole mounts. Computer control systems for microscopes are now widespread, with stages, focus, filters and cameras being controlled from a screen. Electron microscopy is used for higher resolution analysis than is achieved by light microscopes. Epi-illumination with white-light – reflectance contrast – is a valuable way to image RNA and (for animals) DNA *in situ* hybridization where sites are detected by the reflective molecule DAB. (For DNA hybridization to plant cells, too much reflective cell wall debris is invariably present.)

Images taken using a 100× oil immersion lens are very close to the theoretical limitation of resolution of the light microscope so no improvement – increase in data quality – is actually possible. Where higher resolution is required, electron microscopy must be used. Some digital systems used with light microscopes may be more effective and certainly faster at collecting images from very low-light fluorescence, while deblurring or confocal systems are required for some preparations and three-dimensional analysis. However, we have seen many laboratories where 'technological' fixes have been applied, failing to overcome fundamental problems caused by a wrongly set up microscope. Many sales representatives (even from major microscope manufacturers) will offer digital options to compensate for deficiencies in their knowledge of setting up a fluorescence microscope. Such fixes may give visually attractive results, but absolute resolution is sacrificed.

11.2 The epifluorescence microscope and lenses

In situ hybridization signal from the fluorescence protocols presented here is localized by epifluorescence light microscopy, where the specimen is viewed from the same side as the illumination. The typical arrangement of the illumination, groups of lenses (each shown as single lens shapes), specimen and critical filters are shown in *Figure 11.1*. In brief, the light source emits a range of light wavelengths, together appearing white (the spectrum emitted from a mercury vapor lamp is shown in *Figure 11.2*). The light passes through the condenser lenses and then filters. The three key filters are shown: the excitation filter blocks all light wavelengths except those that excite the fluorochrome that is to be visualized (*Figure 11.3*). The light then passes to a specialized filter, the dichroic beamsplitter, which reflects some shorter light wavelengths (those that excite the fluorochrome) on to the specimen, and transmits longer wavelengths (those emitted by the fluorochrome) as it returns from the specimen. The barrier filter then blocks transmission of most wavelengths, allowing those emitted by the fluorochrome to pass (and removing stray reflected light or many wavelengths from autofluorescence), before the light goes to the eyepiece lens and camera or eye (*Figure 11.4*). A valuable book describing the microscope and filters is published free on the Internet by Chroma technology (www.chroma.com) and gives extensive details of optics and filters. Notes about key parts of the system are given below.

If DAPI fluorescence is to be used, the optics must be transparent to UV: lenses must carry a 'fluor' or similar designation and filters between the bulb and specimen must transmit UV. Similarly, immersion oil (or, where appropriate, glycerine) must be a special grade specified for fluorescence microscopy (as well as compatible with the optics, see Section 11.4); other qualities may be either autofluorescent or opaque to UV.

It is essential to have a filter to absorb infra-red light in the light-path; long wavelengths may not be blocked by other filters, can cause heat damage, and will scatter and fluoresce in the microscope (sometimes recording on film or CCD cameras, perhaps as irregular blotches or halos, although not visible by eye). Some microscopes (and particularly epifluorescent modules added as accessories) do not work because of lack of this filter. There are also apertures to minimize scattered light and it is helpful to have neutral density filters (16× is useful) to reduce the excitation intensity (and hence fading) when looking with bright fluorochromes, particularly DAPI.

The remaining parts of the optics of the microscope are standard. For DNA:DNA *in situ* hybridization, the most useful lens is the ×100, giving essentially the highest magnification available from a light microscope, limited by the wavelength of the light. Because the fluorescence is often weak, and the slide has oil from the ×100 lens, it is very helpful to have a high numerical aperture, lower power, oil immersion lens for scanning the slide as well. We find the unusual magnification of ×20–30 is very useful.

Camera systems are discussed below (Section 11.5). The microscope camera system should be such that all the fluorescent light can be directed to the camera, since exposure times may be 1–4 min. Some microscopes have special adjustments to override the normal split (often 60% of light to camera, 40% to eyepiece). It is convenient to have more than one camera port, so that a film and CCD or video camera can be attached together, or different films used.

It may be convenient to have transmitted light illumination, including phase contrast, on a microscope used for fluorescent *in situ* hybridization analysis, but in a laboratory carrying out extensive work, this is unnecessary and can lead to conflicts of use and the need for frequent re-alignment for different purposes. Since this is not always carried out, the result may be sub-optimal, with, for example, phase lenses used for epifluorescence, poor centering of light or

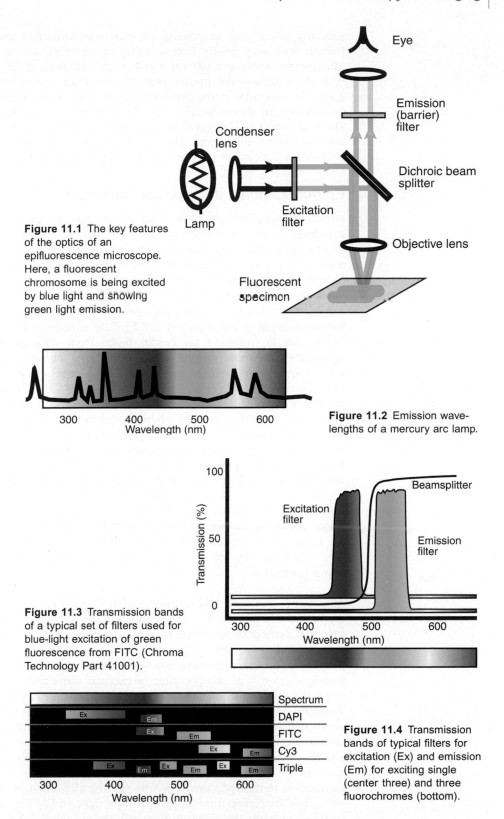

Figure 11.1 The key features of the optics of an epifluorescence microscope. Here, a fluorescent chromosome is being excited by blue light and showing green light emission.

Figure 11.2 Emission wavelengths of a mercury arc lamp.

Figure 11.3 Transmission bands of a typical set of filters used for blue-light excitation of green fluorescence from FITC (Chroma Technology Part 41001).

Figure 11.4 Transmission bands of typical filters for excitation (Ex) and emission (Em) for exciting single (center three) and three fluorochromes (bottom).

apertures, and wrong film types/exposure values selected. Specimens mounted and stained using the fluorescence protocols presented in this book have little contrast after hybridization so transmitted light illumination is not useful.

Microscope hardware

Four large manufacturers make the highest quality epifluorescent microscopes, Leica (formerly Leitz), Nikon, Olympus and Zeiss (see Appendix 2). All are similarly priced; the knowledge of local representatives and after-sales service (which both vary greatly from country to country), are as important as price. Microscopes no longer last for a researcher's lifetime. A 10-year-old microscope is both of technically poorer performance than a new instrument, and in a specialist molecular cytogenetics laboratory will be worn out. Microscope lenses do not have an indefinite life under the conditions used for *in situ* hybridization. The high power UV light will make lenses (or perhaps the glue used in assembly) autofluorescent after about 2000 h of use. The microscope must be sited in a completely dark room (see laboratory design, Chapter 2).

Epifluorescence microscope illumination

For most *in situ* hybridization applications with the fluorochromes discussed in this book, a 100 W mercury vapor lamp (usually size/part code HBO 100) is most suitable. This source is much brighter than the 50 W mercury bulb widely used until the 1980s, and emits light at discrete wavelengths that excite fluorochromes used for *in situ* hybridization (*Figure 11.2*: for some fluorochromes, halogen or xenon lamps, with broad band output, are required). HBO 100 bulbs are expensive (US\$200) and have a relatively short life-time, always specified by the manufacturers and typically 200 h; if the running time is less than specified, bulbs may be returned for credit. Bulbs from different manufacturers have remarkably different properties in brightness, change in performance with time, life-time in practical applications and amount of flicker. The Japanese Ushio bulbs, and a specialty bulb by Osram sold only through Partec (not in the Osram stock lists), are particularly suitable for microscopy applications. Most frequently, bulbs are replaced when they show extreme amounts of flickering or become unusably dim (the quartz glass envelope is visibly dark) but occasionally, bulbs explode (in our laboratory, we keep spare mirrors and condenser lenses, less costly than a bulb, since they may be broken and replacements take weeks to arrive). Hour meters should be fitted to the bulb to record running hours. Turning bulbs on and off reduces their life; after switching on, bulbs should not be switched off for 30 min, when they have reached their full and stable operating temperature. After switching off, bulbs should not be switched on again until cool, taking 30–45 min, and if they are to be used again within 3 h, they should be left on. A new bulb should be left on for at least 2 h before being turned off.

Alignment – centering and focusing – of the fluorescence bulb (and its image from the rear condenser mirror), using the condenser lenses, mirrors and bulb adjustments, is a complex but critical task. If a DAPI image changes in brightness at all as it is moved across the microscope field, then the microscope is not aligned properly or is dirty. Alignment must be repeated after changing bulbs and approximately every 8 h of use (because of thermal movement and drift). Even some company-trained technicians are unable to complete optimum bulb alignment. Typically, there are six, inter-related, knobs and screws for adjustment: some (not necessarily less important) may require unusual tools, while others with identical screws may dissemble the case! Microscope instructions give, with varying clarity, alignment methods, and some Internet sites (e.g. www.zeiss.com/micro/news-tips) give useful advice for all microscopes. The following notes some common approaches to bulb alignment. Before starting alignment, with the lamp switched off, check the connections to the bulb are tight and other adjustment screws are working and near the center of their adjustment range. With a new or badly misaligned bulb, wearing a full-face mask for UV and explosion protection, turn on the bulb with the illuminator case removed from the microscope and held in the hand. Shine the lamp towards a wall or white paper about 1 m away, and adjust the lamp-house screws until the bulb, and its image from the rear lamp-house mirror, is seen to illuminate a circle on the wall uniformly and brightly. (This must be carried out within a few minutes as the case will slowly heat up until it is too hot to handle.) After replacing the lamp-house, it is helpful to remove the objective lens or lens turret and (again with face mask) make the illumination uniform and bright on a sheet of paper placed on the stage; check with all filter sets before replacing the lenses. Finally, carry out minor adjustments with microscope specimens, checking with all filters and lens combinations, while moving specimens across the field of view. Normally, all the adjustment screws will need altering several times.

11.3 Filters for epifluorescence

The three critical filters for selecting wavelengths must be conveniently changed so different fluorochromes can be examined sequentially. In many microscopes, the three filters of each set are arranged in single holder, block or cube (sometimes called Ploemopak by Leica) which slide into place in the light path, while other microscopes contain the filters in single or multiple, interlocked or independent, sliders. Advanced microscopes may place the filters in wheels which sometimes rotate into position under computer control (Speicher *et al.* 1996). One system uses a multi-bandpass excitation filter and dichroic mirror, combined with the emission filters in a wheel, so each fluorochrome is imaged separately with the appropriate emission filter (developed in D. Pinkel's laboratory). Another system under development uses time-resolved fluorescence. After excitation with light, different fluorochromes emit their fluorescence during a characteristic time period. Where the period is long (milliseconds), pulses of excitation light can be synchronized with collecting the fluorescence after the pulse is turned off. This may enable an extra parameter to be detected (effectively giving extra detection 'colors'), and is reported to be valuable in removing specimen autofluorescence, which seems to have a short decay time.

The quality of filters is critical for analysis of fluorescent *in situ* hybridization. Filter technology has improved markedly since the mid-1990s, and specialized filters for most fluorochromes are available with high light transmission and sharp wavelength cut-offs. Filters have a life of 1000–2000 h before surface coatings are damaged by light, start to peel, and become scratched from occasional cleaning. A worthwhile upgrade to older instruments is often new sets of fluorescence filters: modern sets are available for a few hundred dollars.

Three major manufacturers make state-of-the-art fluorescent filters for microscopy and supply them mounted in blocks (cubes) for most microscopes: Chroma Technology, Andover Optical and Omega Optical, all located in New England, USA. Filter sets are also available from the microscope manufacturers, but these are more expensive and seemingly have poorer performance than those from the specialist filter manufacturers. The three specialist filter companies have helpful Internet sites (Appendix 2) that give the spectra of the filters and specific recommendations for each fluorochrome.

Figure 11.3 shows the transmission characteristics of the three filters used for FITC fluorescence. *Figure 11.4* shows the transmission characteristics of other filters. Sets of multiple band-pass filters, where light is transmitted in two, three or four different wavelength bands, are available and a useful *addition* to the single band filters since they enable fluorescence of, say, DAPI, FITC and Cy3 to be examined together. However, they do not replace single band-pass filter sets because they require the brightness of each fluorochrome to be balanced and show greater cross-excitation and spectral overlap of the fluorochromes than single-wavelength filters.

11.4 Fading of specimens and refractive indexes of mountants and oils

Fading of the specimen is a major problem in fluorescence microscopy. Bright light, particularly UV and blue wavelengths which have the highest energy, degrade most fluorochromes and may destroy the specimen by heating. Before reaching the high-performance excitation filter, it is essential that the microscope has a convenient method to block ('turn off', although the bulb itself must not be turned off) the light. The light will also damage lenses and filters, so must also be blocked when the microscope is not being used but the bulb is on.

The fading process is effected by the wavelength and intensity of excitation light, pH, mountant, temperature, oxygen and substances that quench fluorescence. Newer fluorochromes are designed to be more stable. Use of antifade reagents in the mountant between the specimen and coverslip is very important. These may take several hours to become effective, so it is often better to delay photography until the day after mounting a preparation. To reduce fading, observation and photography must be carried out quickly – fading will occur if chromosomes are counted with DAPI staining before photography. For optimum results, photograph or image the longest excitation wavelength first and the shortest last (normally, red emission first, then green and finally DAPI; we take multiple wavelength pictures after single wavelengths).

Using unmatched components in the light-path can systematically degrade resolution and image brightness. The glass used in lenses has a particular refractive index and lenses are designed to work at particular distances from the specimen. The refractive index of immersion oil, coverslip glass, coverslip thickness, mounting medium and mountant thickness must all be compatible and specified for fluorescence applications.

11.5 Electron microscopy

Both transmission and scanning electron microscopy can be used for analysis of *in situ* hybridization results at high resolution (see Section 10.8). The resolution of the light microscope is limited by the wavelength of light, but the magnitudes of shorter wavelength of electrons means technical and specimen considerations limit the resolution. Imaging *in situ* hybridization with the electron microscope is relatively straightforward and requires no particular expertise. There is seldom any need to image material at high magnifications, because the fixation mehods are not optimized for morphological preservation.

11.6 Imaging, microscope cameras and film

Photographic, digital (CCD) and video cameras are fitted to microscopes. A high-quality photographic system is the most important, and when this is optimized, results of high publication standard will be obtained from nearly all experiments (Bracegirdle and Bradbury 1995). Furthermore, the images are of high resolution, archival quality, easily compared from one year to the next, and there is no possibility of 'improvement' to the raw data before storage. Where a microscope is well adjusted, some high-resolution CCD cameras are also suitable for analysis of results from *in situ* hybridization.

Most conventional video cameras are not sensitive enough for fluorescence images, and low-cost video and CCD cameras are usually of low resolution and color depth. Image resolutions and color depth is increasingly quoted in terms based on computer screens: the number of pixels – picture elements – per side of a rectangular picture, and the number of colors possible at each point. A medium resolution computer system will typically have a display size of 1024×768 pixels with 16,000,000 colors (24 bits, 2^{24}, color depth or band width, requiring 2 Mb display memory), while a low resolution system (e.g. found on older portable computers) will show 640×480 pixels in 256 colors (300 kb memory). The resolution of typical consumer types of 35 mm 400 ASA color film is at least 5000×2500 pixels, with a band width of 48 bits or more (representing 2^{48} or 281×10^{12} different colors) – giving an image size of 75 Mb, which is close to (although below) the resolution of the microscope. Even high-resolution digital cameras give only a few percent of this information. While all the 75 Mb of information will not be used during analysis, there is surprisingly often a need to analyze in detail one color (e.g. the red fluorochrome) on one segment of a chromosome, requiring all the information. Furthermore, the film storage is cheap, compact and of archival quality.

Films and film cameras

Film technology has improved markedly through the 1990s, particularly for color print films where there has been the greatest market requirement, with reduction of grain size for given speeds, introduction of shaped grain technology and a wider exposure latitude. There is a considerable difference in films from various manufacturers in microscope applications, with Fuji films generally giving higher contrast and more exposure latitude than Kodak films. Because of the lower cost, ease of reproduction and wider exposure latitude, color print films are generally preferable to slide (reversal) films. For fluorescence microscopy, 400 ISO (ASA) is the optimum speed; the actual exposure times and settings for any microscope must be determined by trial and error, but on a typical microscope with 100 W HBO lamp, will be about 2 s for DAPI images up to 40 s for red images. Automatic exposure meters may be accurate for one filter combination, but not measure the nearly monochromatic light from another filter. To obtain prints with black backgrounds, the automated exposure times on printing machines must be overridden: small developing and printing shops where it is possible to speak to the owner and machine operator will often be willing to do this, especially when shown a sample image. For transmitted light microscopy, 25 or 100 ISO films have higher resolution, and exposure meters are usually accurate.

Black and white photography is usually more expensive than color photography, and obtaining prints with uniform and clean black backgrounds using hand equipment can be very difficult. However, it may be better because contrast can be manipulated more easily and, for transmitted light applications, filters complementary to stain colors can be used.

CCD Cameras

Many manufacturers make CCD cameras connected to computer imaging systems, with cameras and associated software varying in price over a 2000-fold range from $50 (sold as web-cameras or similar). At the high end, systems such as those from Photometrics have resolutions of several million pixels and 24 or more bits of color information. On such systems, the CCD arrays are cooled to reduce background noise, and extremely weak fluorescence signals can therefore be detected reliably by integration (collection) of the signal over several seconds. Many cameras come with dedicated computers and programs that control microscope functions as well, and allow capture and analysis of chromosomal or other images.

Spectral imaging

The 'ultimate' digital camera system of the late 1990s is the spectral imaging system (Liyanage *et al.* 1996; Müller *et al.* 1997; Schröck *et al.* 1996). Rather than measuring the brightness of the fluorescence at each pixel at several wavelengths (defined by the band-pass filters), the system collects a complete spectrum (using a Raman spectroscope), effectively increasing the color depth from 8 bits in each of red, green and blue channels (24 bits), to 96 bits or more. Applied Spectral Imaging (Appendix 2) makes the most widely used system.

Confocal microscopy

Conventional microscopes, whether transmitted light or epifluorescence, illuminate the whole specimen simultaneously and collect light not only from the plane

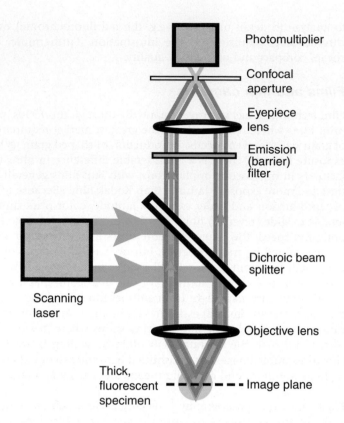

Photomultiplier

Confocal
aperture

Eyepiece
lens

Emission
(barrier)
filter

Dichroic beam
splitter

Scanning
laser

Objective lens

Thick,
fluorescent
specimen
Image plane

Figure 11.5 The key
features of the optics of a
scanning confocal
microscope. Fluorescence
(here, excited by blue light
from the scanning laser)
from above and below
(light paths in orange) the
plane being imaged (light
path in green) is lost at the
confocal pinhole aperture.

of focus, but light from above and below that plane. They are therefore unable to
obtain clear images from within thick sections or whole mounts, or other preparations that are not entirely flat. The confocal microscope avoids these problems
by scanning individual points on the specimen with a spot of light (usually from a
laser), and the fluorescence from each point passes through a pinhole confocal
aperture and is measured by a photomultiplier (*Figure 11.5*; for details see
Sheppard and Shotton 1997). Light from above and below the plane of focus is
not focused through the aperture and therefore makes little contribution to the
brightness measured. The brightness of fluorescence from each pixel is measured
by the photomultiplier, and converted to an image by a computer. In a thick
section or whole mount, each plane can be imaged separately to build up a three-
dimensional representation of the specimen. Alternatively, individual focused
planes can be added to give a flat (squashed) representation of the specimen.
However used, the confocal microscope is convenient to obtain a sharply focused
digital image of a specimen.

New developments of confocal microscope hardware are likely to increase the
resolution considerably (Hell *et al.* 1994).

11.7 Computer equipment

Computer equipment capable of handling full publication resolution digital
images is now available for less than the cost of a 100× oil immersion lens, while
as recently as 1990 only a few printing houses had such equipment. A reasonably
powerful microcomputer (PC or Macintosh), with a large amount of memory
(>128 Mb) and suitable data storage devices are now widely used for analysis of
images from *in situ* hybridization. For example, the contrast and brightness of
color images can be adjusted, overlays of different images (typically DAPI and
different *in situ* hybridization targets) can be made, chromosomes can be cut out
and aligned at any magnification for karyotyping.

Each input and output device has specific color properties, both the types of color used and the brightness of individual colors. The consequence is that an image may look widely different between two different screens, a slide-maker, a dot-matrix printer and a dye-sublimation printer. Differences may be far greater with one sort of image (e.g. fluorescent micrographs of chromosomes) than with another (e.g. of people). Color management software, with calibrated drivers for each device, overcomes this problem, and are supplied with some programs.

Scanners

The image must be in digital format to be processed by the computer. CCD cameras or confocal microscopes will give direct digital images. From films, it is better to scan original negatives (or slides) rather than prints: prints have already been 'processed' in making the print, and large amounts of data are lost compared to the negative. Many photographic shops will offer a service to digitize slides or negatives to the Kodak or Fuji Photo-CD format, costing about $1 per scan, and these are excellent for publication. Alternatively, film scanners with high resolution are available from Nikon or Canon for less than $1000 which give resolution suitable for publication. Other makes may be less suitable although faster.

Printers

The best printers are those based on dye-sublimation or photographic systems such as the Fuji Pictrograph color printers. Their output is indistinguishable from photographic prints, and typical color images that are printed are between 10 and 20 Mb per full-page plate. Recent generation color printers based on dot-matrix systems (e.g. Hewlett-Packard, Epson or Canon) give very high-resolution near-photographic quality images when special glossy papers are used. The running costs of dot-matrix printers are significant: a single print, with black background may cost $5 in ink and paper.

Most color printers also print on overhead projector film for presentations, although the quality is not as good as that of projection slides. Digital slide-makers connected to the computer can be used to make these.

Data storage

In the late 1990s, image storage is a major problem with no long-term, fast, archival and stable hardware available. A typical plate for publication may have 10 images each of 10–15 Mb in size. During preparation of a manuscript, these are stored individually, as composite images, with overlays, intermediate versions (with different layouts, labels, contrast adjustments), and as versions for projection slides. 250 Mb of storage can be used easily for one plate. Tape, magneto-optical and magnetic disk cartridges from only 5 years ago are already almost unavailable, and the hardware drives may no longer be usable on computers with new plug or bus standards, or new operating systems not supported by the original drivers.

Currently, the best solution is the use of writable CD-ROMs. This format can be read worldwide by most recent computers, is compatible with both Macintosh and PC formats, made by several manufacturers and widely enough used that disks will still be readable in a few years, while media is cheap and available. However, the ruggedness and archival stability of the disks is questionable, and they are slow to write.

11.8 Deconvolution and chromosome analysis programs

The confocal microscope, discussed briefly in Section 11.6, does not image significant light from out of the plane of focus. An alternative approach to optical sectioning is the digital subtraction of out-of-focus information from a conventional microscope image. A stack of digital micrographs – typically three to nine – representing a through-focus series, is collected. A computer program then defocuses the images above and below the central image and, using a Fourier transform algorithm, subtracts the data calculated to correspond to the out-of-focus information in the central image, giving a sharply focused optical section of the original material. This method, developed for chromosome applications by David Agard and John Sedat (University of California, San Francisco, CA, USA) is now available commercially.

For analysis of chromosome images, numerous specialized computer programs are available; many are designed for use with human preparations following *in situ* hybridization, but may be flexible enough to use with other species. The programs may be very comprehensive, for example, finding metaphases on the slide, and automatically producing fully analyzed karyotypes from multi-target *in situ* hybridization experiments. A number of specialized systems are available for analysis of the ratio of signals from two or more fluorochromes, as used for CGH (comparative genomic hybridization).

For general image processing, two programs are very widely used for results from *in situ* hybridization: NIH image and Adobe Photoshop, both available for PC and Macintosh platforms. Photoshop is widely used for producing publication images, and has many lookalikes, some free. Photoshop itself is available with 90% academic discount (Appendix 2). NIH image is available free from the web-site ftp://zippy.nimh.nih.gov and has numerous plug-in sub-programs for chromosomal image processing. These include support for quantitative image analysis, and for confocal microscopes. For example, ftp://horus.sara.nl/norbert has the program Object-image (Vischer *et al.* 1994) for making measurements via NIH image. Routines can be written for specific applications.

11.9 Image processing

The normal purpose of image processing for publication in primary journals is to show your data to peers for evaluation. In other cases, further information can be extracted from the images – the quantitative measurements of signal brightness and position, overlapping of different images of one specimen, cutting out and aligning chromosomes, or comparison of results from multiple cells, for example. In analysis of fluorescence images, it is often informative to look at the red, green and blue channels of the image individually: red and green represent the two most often used fluorochromes, while DAPI is green and blue mixed. In this section, a few key points about image processing are discussed; several books cover the topic in detail (e.g. Adler 2000).

Some authors produce photographs that are best described as interpretive diagrams, and bear little relationship to the original image. Therefore, it is essential to avoid using functions to 'paint out' signals considered to be background, or to 'improve' particular hybridization sites in the processing of images for publication. Processes of color balancing, filtering, overlaying of images, or brightness and contrast adjustment, can be used – essentially functions that affect the whole image equally.

A common fault is over-processing of images; any processing reduces the amount of data present, and successive rounds of processing of an image will seriously degrade the amount of data present. The effect can easily be seen by looking at color-level histograms for the image (in Photoshop, using the Image-Adjust-Levels function): if the level distribution is not almost continuous over the whole range, the image is likely to be unsatisfactory. It is important not to need to re-process the brightness and contrast for the final output printer required.

During processing, images must be kept at high resolution. High-compression storage formats that lose data should not be used more than once. The JPEG (Joint

Photographic Experts Group) compression format is very efficient and large, high-resolution images can be compressed to fit on a single floppy disk. However, while a single round of storage using this algorithm will often not be detectable, two or three uses on a single image will be unacceptable. It is worth keeping labels, arrows or numbers in separate layers – late adjustments are often required, and image versions for projection slides will need different indications.

Conversion of color images into black and white for publication is a useful function of the programs. Default conversions may not provide optimum conversion, particularly of red colors. In Photoshop, it is preferable to either use the Image-Adjust options to remove the color information with manual intervention before conversion, or else use the custom actions option available from Photoshop version 5.

CHAPTER 12

Troubleshooting, optimizing results and other frequently asked questions

This chapter will discuss general points about controls and troubleshooting in *in situ* hybridization experiments, with particular detail about problems encountered using fluorescent detection of hybridization to DNA targets. Where possible, we have tried to give suggestions based on experience from our laboratory, although others may have alternative suggestions, and experience gained from different materials and variants of the procedures is a great help to troubleshooting and optimization. In general, 'background' is the greatest problem with *in situ* hybridization experiments, and it is discussed in Section 12.3. Where methods to overcome problems are common to several faults and do not involve following protocols elsewhere in this volume, they are discussed in Section 12.4. Section 12.5 gives suggestions to optimize results from chromosomal DNA *in situ* hybridization experiments.

12.1 Use of controls

For hybridization to cellular and RNA targets, controls are essential for interpretation of the distribution and specificity of hybridization (see Section 10.7 and the box on p. 145). With chromosomal DNA targets, negative controls are of less value for interpretation of results: in a successful experiment, hybridization signal should be seen on both chromatids and at the same position on both homologous chromosomes in multiple cells. However, special controls are needed for methods such as *in situ* primer extension (see Section 8.5). Slides treated following the *in situ* hybridization protocols, but omitting labeled probe from the hybridization mixture ('no probe'), are valuable for troubleshooting problems of autofluorescence and lack of specificity of detection reagents in all experiments.

Inclusion of positive controls in DNA target experiments is useful for troubleshooting and optimization of methods. They also help to avoid major errors such as mix-ups of probe or target species, or interpretation of probe cross-hybridization. The 45S or 5S rDNA sequences and synthetic oligomers to the telomeric sequences are good control probes to use in most experiments. They are robust, a single probe type will work with many species, they can be detected even with suboptimal denaturation times and temperatures of targets, and there is a high copy number at a small number of chromosomal sites. In many species, there are multiple minor sites of 45S or 5S rDNA sequences that provide a useful standard for the sensitivity of detection. To check all stages of the protocol, the control sequence should use the same labeling and detection systems as the probes being tested. In our experiments, we routinely use rDNA and telomere probes in various combinations in double target experiments with the new probes being tested.

12.2 Problems observed with experiments

We have found that the use of high quality preparations is the single most important factor in obtaining good results. Other frequent causes of failure are: (i) incorrect denaturation of preparations; (ii) poor probe labeling; and (iii) use of old or dirty solutions and materials. The probe mixture and detection have to be correct but are routine procedures and generally work. Incorrect microscope set-up is also a frequent problem and must be checked when no or faint images are seen: in the end, the capture of signals for analyzing the results and publications are essential.

As with any laboratory technique, thoughtful observation and noting of incidents at each step concerning each slide are valuable for troubleshooting and improving results in subsequent experiments. Observations from examining chromosome preparations before denaturation, checking labeled probe, looking at the color and nature of tiny centrifuge pellets, noting solution clarity, viscosity or colors, checking the state of the slides, checking temperatures by hand as well as digital instruments, and looking at surface films in washes, will all pick up obvious but easily overlooked reasons for failure of an experiment. If any mistakes have been made during the procedure it is important to note which slides are affected to assist the assessment of results or failures. Not all mistakes prevent the experiment working, but knowledge helps optimize the procedure in later experiments; it is easier if a failed experiment can be explained by a fault in execution rather than by an unknown problem.

Experiments that end in a 'clean' slide, or with perfect material and no signal, can be difficult to troubleshoot. However, it is usually worth taking photographs to be able to ask advice and even small remnants of material and scratch marks give clues. Our experience has also been that hybridizations, initially regarded as not working, may be satisfactory in some areas of the slide, may improve after one or two days as the antifade becomes more effective, may have weak but still useful signal, or may be interpretable in the light of additional data.

In routine experiments, we use double-target hybridization with two probes detected by green and red fluorescence, and counterstain the chromosomes with DAPI to find them easily and to check the morphology. It is very important to examine the entire area of the preparation, as often not all areas might have failed or different problems might affect different areas. Similarly, it is important to compare slides within one run and with previous runs, to deduce the origin of any problem, e.g. if a probe is faulty, only slides with this probe should be affected.

No or faint fluorescence with some or all filter combinations

This problem is most often connected with the setup of the microscope, the mountant or oil used for viewing.

1. Check the setup of the fluorescence microscope. Make sure the light path is not obstructed, and that the mercury lamp is not too old, and gives uniform, stable light.
2. If red and green fluorescence is seen, but DAPI fluorescence is not seen or very weak, the optics, oils or mountant might not allow UV light through to excite DAPI. Possibly, DAPI staining did not work, or was omitted (*Protocol 9.3*).
3. If the microscope immersion oil becomes mixed with mountant, water (moisture in the air during storage) or different makes of oils, light will not be transmitted and the images become poor and may show continuous movement: remove coverslip, wash slide in detection buffer, and remount.
4. If the mountant does not contain an effective antifading reagent, signals may disappear in less than a second (see Section 11.4). It may take 24–48 h for antifade solutions to become fully effective. Also if slides have been stored for

a long time, antifade mounting solutions can become ineffective: again, remove coverslip, wash slide in detection buffer, and remount.

Nothing or only little specimen material present on slide

Loss of material frequently reflects incorrect slide or material preparation.

1. Is the preparation there? Are any chromosomes and nuclei present? Are whole sections lost? If nothing is present, the slide might be upside down (no preparation on top), or the preparation may have been physically removed at some point (by wiping the wrong side or scraping with another slide).
2. Is anything visible with through-light or phase contrast observation? If so, then staining or microscope set-up may be incorrect. (See 'No or faint fluorescence' point 1 above, Section 12.4 and Chapter 11.) Check counterstaining (*Protocol 9.3*).
3. Often, some material may be present, but selective loss, particularly of metaphases in spreads or nuclei from sections, has occurred. This is most likely due to incomplete or no pretreatment of glass slide with acid (*Protocol 5.1*) or coating of the slide (*Protocol 6.2*), too high denaturation temperature, poor preparation, or use of water rather than SSC/PBS solutions or alcohol for washing: improve preparation methods, and optimize denaturation (Section 12.5). Specimens may have dried out during hybridization.

Preparation present, morphology good, but no signal seen

If no signal is seen, first make a thorough check that this is really the case: sit in an entirely dark microscope room for 10 min before observation, and check the entire area of preparation on the slide. It is essential to check whether there are some chromosomes, nuclei or cells present somewhere on the slide that show some signal to indicate the problem. Causes include:

1. Wrong side of slide treated with hybridization or detection solutions. In this case, chromosomes and nuclei usually have many scratches.
2. Wrong detection system used (e.g. anti-digoxigenin for biotin) or wrong counterstain used (propidium iodide with red detection).
3. Probe labeling is poor: check another batch of probe and check labeling of probe with a dot-blot (*Protocol 4.7*). See Section 12.5 for optimizing results.
4. Probe has no homology to preparation: use a positive control where the probe should hybridize.
5. Detection reagents have failed or were used at too low concentration: examine positive controls, e.g. with 45S rDNA.
6. Antifade not effective: see 'No or faint fluorescence' above.

Preparation present, morphology good, very little signal seen

If some green or red fluorescence is seen, this normally indicates that the detection reagents are working. Controls (see Section 12.1) are very helpful for troubleshooting 'no signal' results. Consider the following causes:

1. Probe labeling poor: check another batch and check labeling (*Protocol 4.7*). See Section 12.5 for optimizing results.
2. Denaturation temperature too low: increase temperature, typically by 3–5°C.
3. Hybridization mixture incorrect: check and adjust (Chapter 8, *Protocols 8.2* and *10.2*).
4. Probe has no homology to preparation: use a positive control where the probe should hybridize.
5. Washing incorrect or solutions contaminated: check and adjust (Chapters 9 and 10).
6. Drips of water on slide during hybridization (raising stringency, possibly to

more than 100% so there is no hybridization; see Chapter 7): adjust hybridization oven conditions.

Preparation present, morphology poor, with or without in situ *hybridization signal*

Because of the elevated temperatures and extended times in various solutions, preparation or chromosome morphology is always poorer after the *in situ* hybridization experiment: despite ones hopes, it never improves. Examine the preparation and compare with notes about the preparation and its general appearance with other preparations from the same material following mounting and staining (*Protocol 9.3*) without hybridization steps. Phase contrast micrographs taken before the *in situ* hybridization are useful both for making accurate chromosome measurements and troubleshooting.

1. If preparation is poor but similar to original, use improved preparation methods (Chapters 5 and 6) to optimize preparations. Re-fixation or pretreatments with proteases or other solutions (*Protocols 8.1* and *10.1*) may also improve hybridization, although results are not as good as with optimal preparations.
2. If chromosome morphology is lost (the chromosomes have become ghost-like or look like inflated balloons and cannot be viewed in one focal plane) and the hybridization signal from genomic or painting probes is stronger at the edges and heterochromatin of the chromosomes, they are overdenatured (*Figure 12.1a, b* and *d*). This can be caused by either too high or too long denaturation or by too strong digestion with protease: reduce the time (typically by 2^{-5} min) and temperature (typically by 3–5°C) of the denaturation, and reduce the concentration of protease (typically by 50%).
3. If chromosomes are lost selectively from the slide, the slides may not have been pretreated (*Protocols 5.1* and *6.2*) or re-fixed after preparation (*Protocols 8.1* and *10.1*).

Scratches

Scratches on preparations are easily introduced by careless handling of slides. They are seen as lines running across slides involving loss of material. Particular causes are slides touching each other in staining jars, or edges of coverslips touching material during removal.

12.3 Background

Almost any hybridization would be improved by reduction of the background strength, and background is the major limitation to sensitivity, although some background will not prevent analysis. There is a range of types of background depending on the cause; analysis of background can give valuable clues for troubleshooting and improving the technique.

Carefully examine whether the background is visible with all filter sets. What shape, size and color is the background? Where is it? Is it over chromosomes and nuclei, over cytoplasm, or everywhere? If signal is on chromosomes, does it occur on the same chromosome in the same position in each cell?

No specific signal – preparations glow

Where there is a uniform and general glow from the nuclei, perhaps with occasional patches elsewhere, the result suggests that there is no label molecule present for detection (probe or hybridization is faulty). If the glow is very intense with occasional hybridization signal, the detection reagent concentration may be much too high (e.g. 10×): reduce strength in another experiment. If background fluorescence is seen over the whole preparation, and is not focused but sometimes in

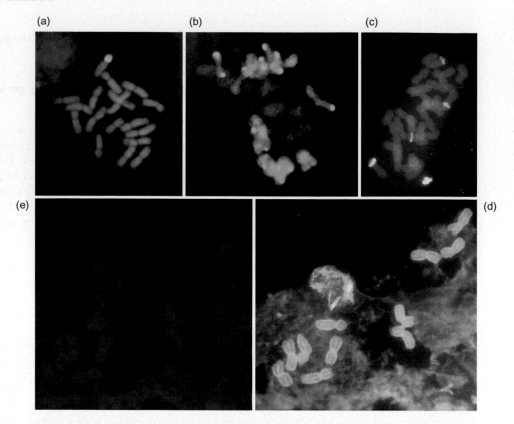

Figure 12.1 Problems with *in situ* hybridization. (**a**) Part of a metaphase of a tetraploid wheat species hybridized with the 45S rDNA sequence. Chromosomes are slightly overdenatured, and only major sites are visible. A large particle of fluorescent material is present (arrow; note the color of the label and angular edges). (**b**) A meiotic metaphase I of an intergeneric hybrid labeled with total genomic DNA from one of the parental genomes. Chromosomes are overdenatured and terminal heterochromatin has labeled more strongly. (**c**) Genomic *in situ* hybridization to part of a wheat metaphase with a *Hordeum chilense* chromosome arm. Too much probe has been used, so all chromosomes show fluorescence, although there is still adequate discrimination of the alien chromosome segment. (**d and e**) Part of an overdenatured barley metaphase with too much cytoplasm. No signal is seen with either green or red fluorescence. In (**d**), peripheral labeling of chromosomes suggests overdenaturation, while the overall red color (**e**) suggests too high a detection reagent concentration.

patches, the problem may be autofluorescence of microscope lenses, mounting media, immersion oil, or material underneath the condenser (see Section 12.2). After storage, fluorescence may become very faint and high overall background fluorescence can be observed. Exchanging the mounting medium can sometimes revitalize old *in situ* hybridization preparations.

Cytoplasm or other material on the surface of the preparations

Cytoplasm normally does not fluoresce strongly with DAPI, so chromosomes and nuclei might look satisfactory when viewed with UV. Under blue excitation, cytoplasm may appear milky yellow-green but with green excitation it is often less visible, or appears as irregular red areas. Cytoplasm will surround nuclei and chromosomes, with black areas between, while 'films' and other material on the surface of the preparation will cover everything and show a high uniform background fluorescence with scratches and patches. Cytoplasm masks chromosomes and nuclei, such that reagents cannot penetrate and denaturation is not effective, failing the hybridization and detection. Further dirt and unspecific hybridization and staining are attracted and cause high background (*Figure 12.1d* and *e*). Films on the glass slides can come from insufficient cleaning of the slides or can be introduced during washing steps. As suggested previously, carefully scan slides to see if there are any cells free of the cytoplasm or film. Improving the slide and chromosome preparation methods (Chapters 5 and 6), even starting with new fixations with a high metaphase index, are normally the best answer to problems with cytoplasm and background; be very selective in choosing slides suitable for *in situ* hybridization. Refixation or pretreatments with proteases or other solutions (*Protocols 8.1* and *10.1*) may also improve hybridization, although results are not as good as with optimal preparations (see Section 12.2).

Large fluorescent or black dots and crystals present

Dirt and crystals are accumulated during chromosome preparation and during the *in situ* hybridization procedures, by using dirty cover slips, storing slides in the open before *in situ* hybridization, letting the slides dry out between steps, by picking up films on the surface of solutions or by using contaminated solutions. Dirt particles, remnants of cell walls and crystals normally fluoresce strongly with all excitation wavelengths, show variable colors, often have sharp edges and are distributed randomly over the slide (*Figure 12.1a*). Washing solutions contaminated with bacteria can ruin hybridization, and bacteria, fluorescing blue with DAPI and looking like a small chromosome (*Figure 12.2a*), sometimes can be seen on slides. Normally, with familiarity and confidence with the procedure, dirt will be reduced automatically, but it is advisable to work cleanly and carefully at all times, and to check solutions regularly for contamination.

Cross-hybridization of probe

Probe deposition may occur unspecifically to low- or non-homologous DNA targets and to non-DNA targets, appearing as signal dots that may be the same size and color as signal sites. Cross-hybridization to non- or low-homologous sequences will only appear on nuclei and chromosomes (at defined positions), while non-DNA targets are most often cytoplasmic or chromosomal proteins (more random sites; *Figure 12.3e* and *f*). Background distributed randomly over the whole slide is more likely caused by detection or probe problems (see 'No specific signal' on p. 161 and 'speckles' on p. 164). When the target for the probe is not available because of masking by cytoplasm, incorrect denaturation, use of wrong probe or wrong stringency, nonspecific hybridization will be increased. Normally, high amounts of blocking DNA from salmon sperm or *Escherichia coli* (Sections 8.3 and 10.2) are included in the hybridization mix, or preparations are acetylated (*Protocol 10.2*) to reduce electrostatic and other nonspecific binding of the probe to nontargets. Further suggestions to overcome the problem include changing the amount or source of blocking DNA, reducing of cytoplasm on the slides, or pretreatment of preparations.

If cross-hybridization to chromosomal sites is seen, adjust probe concentrations, stringency and washing conditions: in particular, careful adjustment of probe and

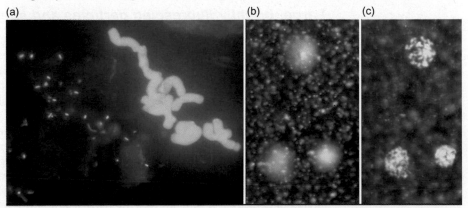

(a) (b) (c)

Figure 12.2 Problems seen with DAPI staining. (**a**) Contamination of a preparation of rye chromosomes with bacteria deposited from the SSC washing solutions (left of field). (**b and c**) Particles of DAPI-staining material in the mountant photographed at two different focal planes.

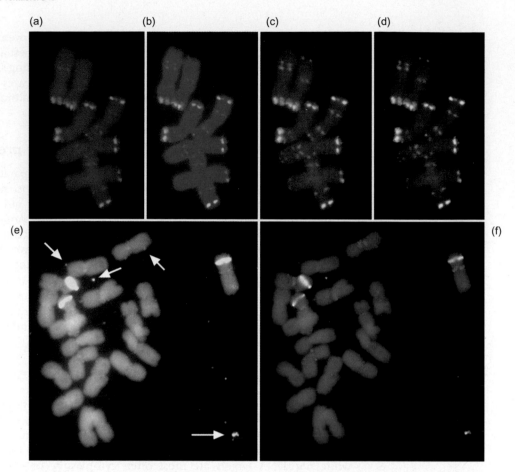

Figure 12.3 Optimization of counterstaining and probe contrast using image analysis. (**a–d**) Part of a rye metaphase with different balances of propidium iodide (red) to hybridization signal (green, or yellow where it overlays red). Minor sites are obscured where the red counterstain is too bright. In this case, good contrast could be achieved by processing (**d**), but in some situations, the counterstain must be washed out before re-photographing. (**e and f**) Part of a tetraploid wheat metaphase probed with the 45S rDNA (yellow) and a simple sequence repeat (red). Some background sites are visible (e.g. arrows), but discrimination of background and real hybridization sites is difficult at some sites.

blocking DNA concentrations is required for genomic *in situ* hybridization, CGH and chromosome painting methods (*Figure 12.1c*; see 'Optimizing probe concentration' Section 12.5).

Large, brightly fluorescing particles and speckles

The presence of multiple, very bright star-like particles and speckles randomly distributed over the slide is a frequent but undesirable result. Where the color is that of the detection fluorochromes, speckles may be caused by disassociation and/or agglutination of the hapten or fluorophores conjugated to the DNA probe or detection reagents. Deduce the cause by comparing slides from one run, or check detection reagent by applying to a control slide that was hybridized without probe. If the detection reagent is at fault, centrifuge briefly and use the supernatant; alternatively, pass through a spin-column to remove agglutinates. The labeled probe may be too long (typically, 800 bp or longer; check length by gel electrophoresis, *Protocol 3.3*), may not be purified from unincorporated label molecule (purify by ethanol precipitation, *Protocol 4.6*) or has degraded in storage (make new probe). If blue fluorescing particles of DAPI are seen (often moving in the mountant; *Figure 12.2b* and *c*), the stain can be filtered through an 0.22 μm filter and the slide restained or remounted after washing in detection buffer (*Protocol 9.2*).

12.4 Troubleshooting

The microscope setup needs to be checked

Microscope setup is vital and is discussed in Sections 11.2 and 11.3. In particular, check the bulb is centered correctly (e.g. some lamp houses have adjustment screws that can be moved unintentionally when covering up the microscope after use). Check that filters, mirrors or irises for specific applications (e.g. polarizers or switching to a second outlet port) do not obstruct the light. Make sure the excitation and analyzing filters are correctly mounted and are specified for the spectra of the fluorophores used (see *Table 4.1* and Sections 11.2 and 11.3). If the DAPI image is weak or very diffuse, check that fluorescence objective lenses and filters are used, and no filters block UV light from the bulb. Immersion oils and mountants must all be specified for fluorescence microscopy use. Age and heating (e.g. in a car) may degrade oils and mountants.

The immersion oil causes problems

Generally, use a very small amount of immersion oil, and be careful to keep it away from the edge of the coverslip. The coverslip should always be substantially larger than the area of the specimen, particularly since the best cells or nuclei are frequently found towards the edge of the original preparation. We recommend making preparations over a 18×18 mm area, carrying out hybridization under 22×22 mm and mounting under a 24×40 mm coverslip. We normally store slides horizontally and leave oil on, but care needs to be taken that oil does not run off the side or take up moisture. If oil has been contaminated, carefully hold the coverslip at one end and wipe off oil with a tissue; reapply new oil.

The mounting medium needs to be exchanged

Hold coverslip firmly and carefully wipe off immersion oil; place slides into 37°C incubator for a few minutes. Lift off coverslip slowly but steadily, and wash slides in detection buffer twice for 5 min. If DAPI stain was very weak, restain the preparation (*Protocol 9.3*) and mount in new antifade mounting medium.

Probe labeling did not work

We test our probes routinely (*Protocol 4.7*) after labeling and discard those not giving satisfactory signal. In most cases, the test is a good indication of how well they will work although a successful *in situ* hybridization experiment is the final proof, and is certainly the best way to optimize probe concentration for maximum signal and minimum background (see Section 12.5). Probes can also disintegrate or the hapten disassociate during storage (repeated freeze-thawing) or heating for probe denaturation.

If the labeling has not worked, we suggest using a different labeling protocol (e.g. random primer rather than nick translation; *Protocols 4.1–4.5*) or different label (e.g. biotin rather than digoxigenin; Section 4.1 and *Table 4.1*) rather than trying to work out why one has not worked. Making a particular method work can be time consuming and expensive in reagents. Some probes seem difficult to amplify by PCR (use an alternative protocol) and some labels (particularly digoxigenin) may be incorporated poorly. Some impurities in DNA used for labeling will inhibit one labeling enzyme but not another. DNA used for labeling can be repurified (phenol-chloroform extraction or commercial purification kit); it may also be worth trying another supplier or batch of labeling enzyme.

Patchiness of hybridization

Uneven spreading and cytoplasm covering the target accounts for much of the patchiness of results seen on many slides: improve preparation quality (Chapters 5 and 6). Other causes are bubbles forming under the coverslip during denaturation, hybridization or detection, or the hybridization stringency may have varied over the slide area because the hybridization mixture has dried out around the edge of the coverslip or accumulated moisture in this area during hybridization.

12.5 Optimizing results

The results from any experiment allow suggestions about improvements: in particular, we will often alter probe concentrations and denaturation conditions, proteolytic pretreatment and sometimes stringency and detection reagent concentrations, with new batches of probe or fixations. Normally, two or three experiments are required to optimize probe concentration and denaturation conditions for a particular probe-target combination.

Optimizing probe and blocking concentrations

Apparent cross-hybridization, too strong or weak hybridization

Probe concentration should be adjusted to optimize the hybridization signal. If signal sites are strong, but chromosomes show background hybridization, reduce probe concentration; increase concentration if hybridization sites are weak (and perhaps not always on both chromatids). Typically, probe concentration is altered by 50% between experiments to give the best signal to background ratio; each labeling reaction will have a different optimum amount. Hybridization time can also be increased if hybridization is not as uniform as required (perhaps 72 h or more), or immunological amplification could be tried (Section 9.4), particularly with low and single-copy targets. If chromosomes show uniform fluorescence, try reducing detection reagent concentrations, or increasing concentrations if chromosomes show no fluorescent background at all.

Genomic in situ *hybridization*

Total genomic DNA may show cross-hybridization to all chromosomes in a hybrid or derived line (*Figure 12.1c*). If hybridization is very strong or not specific, either reduce the probe concentration (by 50% if the experiment is difficult to interpret; to optimize results, changes of as little as 10% are useful) or increase the concentration of blocking DNA (see *Table 8.2*, Section 3.2 and *Protocol 8.2* for sections). If hybridization is weak and not specific, increase the probe concentration and alter the probe to block ratio. Preannealing of the blocking DNA with either the probe DNA or with the chromosomal DNA on the slides can also enhance differentiation. Genomic DNA will not label chromosomes specifically in all hybrid combinations (see Heslop-Harrison and Schwarzacher 1996).

Optimizing stringency

Stringency should be adjusted to limit cross-hybridization to undesired target sequences (see Chapter 7) while giving the highest amount of probe binding (strongest signal) at desired sites. However, alteration of probe concentration may be more important to optimizing results than altering stringency. In general, for probes known to be homologous to target sites in the cells, hybridization should be carried out at a high stringency so that results are as specific as possible. Probes that are members of sequence families or are from heterologous species are usually hybridized at lower stringencies.

Optimizing denaturation

Chromosome morphology should be good: if chromosomes are ghost-like or not intact, denaturation is probably at too high a temperature. Weak hybridization may indicate too low a temperature or time for denaturation. We have found that chromosomes from different species, different tissue, at different stages of the cell cycle, after different time in fixation, surrounded by more or less cytoplasm or prepared by different researchers, need different denaturation conditions. Also, denaturation might differ for the detection of tandemly repeated sequences, tending to need longer times than dispersed or single copy sequences. Section 8.4 and *Protocol 8.4* make suggestions for optimizing denaturation temperatures.

Optimizing counterstaining

The strength of DAPI staining after *in situ* hybridization varies depending on species, age of fixation, preparation method and protocol followed. However, the hybridization signal does not seem to correlate with the strength of counterstaining, and chromosome banding revealed by DAPI after the *in situ* hybridization protocol can be very variable. Check concentration of stains, and increase or decrease if necessary (*Protocol 9.3*) in future experiments. Slides can be completely restained (*Protocol 9.3*, Note 2) or extra stain can be added under coverslips.

Patchy counterstaining is also found occasionally, sometimes due to cytoplasm or other problems in the preparation. Incorrect denaturation of chromosomes may dissolve much of the DNA, giving weak staining. After observation of slides in the microscope, illuminated areas will show fading.

Counterstaining can be too strong, obscuring chromosome morphology and hybridization signal, or showing significant fluorescence in other wavelengths (*Figure 12.3a–d*). Propidium iodide (PI) is often a problem and will obscure weak fluorescein (FITC) signals. Removing the coverslip and soaking the slide in detection buffer (*Protocol 9.3*, Note 2) can reduce counterstaining by PI and, to some extent, DAPI. Image processing, particularly with balancing of the brightness of different color channels, is valuable for optimizing contrast between different fluorophores (*Figure 12.3*).

Optimizing preparations

As indicated elsewhere, optimization of preparations is perhaps the most important factor in successful *in situ* hybridization experiments. Depending on the results required, preparations for DNA:DNA *in situ* hybridization must be well spread and clear from cytoplasm. Chapters 5 and 6 give suggestions for improvement of preparations.

12.6 Frequently asked questions (FAQs)

Questions that we have been asked about *in situ* hybridization, not directly addressed in the rest of this book, are answered below based on our experience. Answers will be updated on our website (search from http://www.jic.bbsrc.ac.uk) and many other tips are available on the web (see Chapter 13) and from Internet discussion groups.

How quantitative is ISH and how large are sites?

On uniformly spread chromosome preparations, using digital analysis it is possible to compare the signal strengths of different sites within one cell (e.g. Leitch and Heslop-Harrison 1992), and semi-quantitative comparisons of signal strengths on single metaphases can be made almost always (e.g. 'major and minor sites').

DNA-target signals appearing as separate fluorescent dots on each chromatid of a metaphase chromosome typically represent a target site less than 100–250 kb long. Larger sites normally appear as bands across both chromatids. Using fiber *in situ* hybridization, length measurements of targets (or distances between pairs of targets) are accurate to approximately 20%, depending on the number of fibers aligned and averaged for analysis.

Do I need to use metaphase preparations? Can I use easy-to-obtain interphases?

The possibility of analyzing chromosome numbers or karyotype rearrangements by hybridization to interphase nuclei rather than metaphases is widely cited as a use of *in situ* hybridization ('interphase cytogenetics'). However, in practice, interpretation of results from interphases is difficult for many reasons: cross-hybridization of probes and background are much more difficult to interpret than on metaphases, and a proportion of interphase nuclei are aneuploid. Although interphase signals can be interpreted, data from metaphases for most questions are much more reliable. Despite considerable commercial research, *in situ* hybridization to interphase nuclei is not yet used routinely to diagnose human trisomies in fetal cells, the alternative but much longer method of culturing the cells to obtain metaphases being preferred.

How sensitive is in situ *hybridization?*

Hybridization of probes less than 1 kb long is routinely detected on human and mouse chromosomes, and companies (e.g. Genome Systems, Appendix 2) offer *in situ* hybridization as a service. On plant chromosomes, targets of 1 kb or less can also be detected, but with less certainty. Using very long times for enzymatic detection, a few molecules of target RNA can be detected.

What is the linear resolution of in situ *hybridization?*

On metaphase chromosomes, two DNA hybridization sites that are approximately 3 Mb apart will normally be clearly separated, and red and green detected probes will be unequivocally ordered; use of extended prophase chromosomes or pachytene chromosome spreads (*Protocol 5.6*) can increase resolution to approximately 500 kb. For sites closer than these distances, red and green labels will overlap, giving a yellow signal. On interphase nuclei, sites down to a few kilobases apart can be distinguished (Brandriff *et al.* 1991), although ordering with respect to the centromere or telomere may be more difficult.

Using fiber *in situ* hybridization, target lengths of a few hundred bases can be detected, particularly when one target is adjacent to a longer, labeled, target.

Why use in situ *rather than Southern hybridization?*

In situ hybridization gives the chromosomal localization of a probe and hence is able to detect multiple sites of a repeated probe that cannot be visualized on Southern hybridization because of overlaying bands (e.g. Maluzsynska and Heslop-Harrison 1990). Often, *in situ* hybridization is the best or only means to determine whether a probe is organized in clusters or dispersed within the genome, what the long range organization of repeated sequence families to each other is, and how probes are ordered along chromosomes. Because single events are analyzed and DNA is packed in chromatin, *in situ* hybridization is not as sensitive in detecting single copy sequences and not as quantitative as Southern hybridization (see above).

Can I use a negative result?

It is very difficult to be sure that a given probe does not exist in the genome. Similarly, a positive result does not guarantee that the target is present in the genome: it means that sequences with normally 80–90% homology to the probe are present at a high enough copy number at a particular chromosomal location to be detected.

How do I report hybridization results in a journal?

Normally, examples of the results are illustrated and text describes the findings from more preparations in general terms; for chromosomal targets, a karyotype (idiogram) is often used to illustrate the distribution of hybridization sites. The detection of hybridization should be reported as showing that sequences homologous to the probe are present (not 'showing that a probe is present'; there may be individual bases that do not hybridize, or the result may come from hybridization of only a fragment of the probe).

How many metaphases should be examined for interpretation?

The detail of analysis reflects the importance of the experiment and material. With material that is very hard to obtain, a single preparation or single metaphase may be significant enough for an article. Analysis of single cells or preparation for each accession may also be enough for surveys, particularly when there is an another experiment (using, for example, Southern hybridization or segregation) to support the result. For analysis of locations of signals (e.g. localizing single-copy hybridization sites, detecting translocations) or generation of karyotypes (using repetitive sequence hybridization), measurement of a minimum of three to five complete metaphases is usually enough. It is useful to confirm data with a number of additional metaphases or prophases (with extended chromosomes and perhaps less complete or with overlaps).

What is the minimum information needed in materials and methods of a published paper?

The composition and temperature of the most stringent steps of hybridization (usually the wash and hybridization buffers) are essential in any paper. Unfortunately, many authors skip this information while reporting routine information about methods of labeling and the detection reagents (these do not affect the result and are secondary to stringency data).

Can in situ hybridization be used for screening of material?

The *in situ* hybridization methods are labor intensive (chromosome preparation is hard to automate), and expensive compared to DNA or karyotype-based tests; however, they can give much more information and can be more sensitive. Hence, they are useful for specialized situations: examining a small number of plant lines of great interest where there is little background knowledge (e.g. wheat lines with a probable small rye chromosome introgression on an unknown chromosome; Ribeiro-Carvalho *et al.* 1997), or for prenatal diagnosis of chromosome abnormalities where there is a family history or where an affected sibling has been born.

How long does it take to obtain publishable results?

This frequent and important question is difficult to address! Given good probes and expertise in making chromosome preparations for a particular species, to add data to existing knowledge, it will take typically 1–2 months for somebody to obtain optimized and publishable results (i.e. two to six 'runs', see timing in Section 1.4, perhaps with additional time for setting up, analysis and preparing illustrations for publication).

With plant material, our current experience is that one person with PhD-level experience, given six independent clones of about 1–2.5 kb long, will be able to localize two with high accuracy in a 2-month project. Two probes will probably not be located, while two will give multiple locations; considerable effort will be required to sort out these, including detailed checks of the nature of the probe, its content of repetitive sequences and behavior in Southern hybridization to multiple restriction digests with very long exposures. One way to increasing sensitivity is to isolate large clones, e.g. BACs, that contain the probe of interest and using these large clones for localization (Gomez *et al.* 1997; Jiang *et al.* 1995). However, total genomic or $C_o t$ fraction DNA needs to be included to suppress cross-hybridization of repeats present in the large clone and, in some cases, even with blocking, a selected larger clone might not give results.

How long can slides be stored?

Slides from *in situ* hybridization experiments usually have a fluid mount and unstable fluorochrome stains. They can be stored flat in slide trays in the dark in a cold room or refrigerator, and will slowly deteriorate; however, results can often still be seen 2 years or more after the hybridization. Removing coverslips, washing in detection buffer and remounting may revitalize old slides.

The enzymes I use for chromosome preparations are no longer available. What can I use instead?

Individual laboratories need to change enzymes and reoptimize mixtures and digestion times when suppliers change enzymes. Crude enzymes, such as those used for degrading cell walls, are manufactured in bulk as impure preparations (for use in paper, chemical and food industries) and sold to laboratory chemical suppliers in large drums. When the drum runs out, another is purchased with different properties; laboratory suppliers may or may not change the catalog number or description when this happens.

Why has everything stopped working?

Even in major laboratories, experiments sometimes stop working for everybody without any predictable reason. A change in water quality is a frequent cause: for example, new purification chemicals (e.g. organic amines rather than chlorine) may be introduced by water companies and not removed by a laboratory purification method, or new pipework may introduce chemicals into supplies. A whole batch of reagent (e.g. antidigoxigenin) from a company may not work (so replacing the chemical may not help): keep careful records of batch numbers, complain multiple times to the company and try other suppliers. Glass slide manufacturing methods may change (e.g. sandblasting rather than acid etching may leave glass particles that prevent direct pressure from the coverslip on the preparation) so preparations are no longer suitable for hybridization. Careful record-keeping and observation will help overcome such frustrations.

CHAPTER 13

Reference sources

Major journals carry the latest findings and methods in molecular cytogenetics – *Nature, Science, Cell* and many others. Review journals, such as the *Trends* and *Current Opinion* groups, as well as publishing reviews of results from molecular cytogenetics, often publish significant technical developments. Primary publications in refereed journals and conference proceedings often give very limited space to the 'Materials and Methods', so once established, an important technique will be included in edited chapters of technique books. BIOS publish authoritative technique books on many aspects of molecular biology. The Oxford/IRL Press *Practical Approach* series is well respected, and the Humana *Methods in Molecular Biology Series* (e.g. Isaac 1994) is good value, while there are also long-running series published by Springer. Many individual books are also published which include chapters about both DNA and RNA *in situ* hybridization experiments.

Manufacturers' catalogues and protocol sheets are valuable reference sources: even if a kit is not purchased, the protocols are usually available on request. Many manufacturers produce helpful and free magazines with protocols, tips, trouble shooting and other notes covering their products in the most rapidly developing areas of molecular cytogenetics research: it is worthwhile joining these mailing lists. There are also a number of free magazines carrying advertising and technical notes – *BioTechniques* is a well-known international publication.

The Internet is an increasingly valuable source of protocols, materials and methods, and advice for *in situ* hybridization. Both research laboratory and teaching protocols are available from many institutions, while the major organisms used in research have dedicated sites, sometimes with protocols and up-to-date citations including *in situ* hybridization experiments. All companies also maintain their own sites, often with product guidance and alternatives. The Chromosome-watch site coordinated by Dr Shigeki Nakayama, http://www.chromosome.net, includes links to major laboratory, organism and corporate sites, while the authors' laboratory web-site, searched from http://www.jic.bbsrc.ac.uk, includes material related to this book and updates to the frequently asked questions and trouble-shooting. Review journals and the journal *Chromosome Research* have regular articles about related web-sites. For particular queries and protocols, the popular search engines usually find sites with useful information: Yahoo, AltaVista.digital, HotBot, Infoseek, Lycos, WebCrawler, Spot, NorthernLight, Excite and UKMax (access with http://www.[NAME].com) are major sites, although new ones appear frequently. The huge, freely accessible website, Geocities, is mostly not indexed by these services, but has useful research information, particularly from students, and it includes its own search engine (http://www.geocities.com).

There are many conferences where the latest results and methods of molecular cytogenetics are included; some have a bias to clinical topics. Conferences on chromosomes, genomes and genomics will include molecular cytogenetics sessions. The Gordon Conference on Molecular Cytogenetics is held in even-numbered years, while the European Cytogeneticists Association meets in odd-numbered years.

13.1 Literature cited

Adler J (2000) *Image Analysis*. Oxford: BIOS Scientific Publishers (in press).

Albini SM, Jones GH (1987) Synaptonemal complex spreading in *Allium cepa* and *A. fistulosum*. I. The initiation and sequence of pairing. *Chromosoma* **95**: 324–338.

Albini SM, Schwarzacher T (1992) *In situ* localization of two repetitive DNA sequences to surface-spread pachytene chromosomes of rye. *Genome* **35**: 551–559.

Ambros PF, Matzke MA, Matzke AJM (1986) Detection of a 17 kb unique sequence (T-DNA) in plant chromosomes by *in situ* hybridization. *Chromosoma* **94**: 11–18.

Anamthawat-Jónsson K, Schwarzacher T, Leitch AR, Bennett MD, Heslop-Harrison JS (1990) Discrimination between closely related Triticeae species using genomic DNA as a probe. *Theor. Appl. Genet.* **79**: 721–728.

Anamthawat-Jónsson K, Schwarzacher T, Heslop-Harrison JS (1993). Behavior of parental genomes in the hybrid *Hordeum vulgare* × *H. bulbosum*. *J. Heredity* **84**: 78–82.

Anamthawat-Jónsson K, Heslop-Harrison JS, Schwarzacher T (1996) Genomic *in situ* hybridization for whole chromosome and genome analysis. In: Clark M, ed. In situ *Hybridization: A Laboratory Companion*. London: Chapman & Hall, pp. 1–23.

Arnoldus EPJ, Noordermeer IA, Peters ACB, Voormolen JHC, Bots GTAM, Raap AK, van der Ploeg M (1991) Interphase cytogenetics of brain tumors. *Genes Chromosomes Cancer* **3**: 101–107.

Bagasra O, Hansen J (1997) In situ *PCR Techniques*. New York: Wiley-Liss.

Bardsley D, Cuadrado A, Jack P, Harrison G, Castilho A, Heslop-Harrison JS (1999) Chromosome markers in the tetraploid wheat *Aegilops ventricosa* analysed by *in situ* hybridization. *Theor. Appl. Gent.* **99**: 300–304.

Beynon RJ, Easterby JS (1996) *Buffer Solutions: The Basics*. Oxford: BIOS Scientific Publishers.

Binder M (1992) *In situ* hybridization at the electron microscope level. In: Wilkinson DG, ed. In Situ *Hybridization: A Practical Approach*. Oxford: IRL Press, pp. 105–120.

Binder M, Roth J, Gehring WJ (1986) *In situ* hybridization at the electron microscope level: localization of transcripts on ultrathin sections of lowicryl K4M-embedded tissue using biotinylated probes and protein A-Gold complexes. *J. Cell Biol.* **102**: 1646–1653.

Blackman RL, Spence JM, Testa JM, Raedy PD (1998) A 169-base pair tandem repeat DNA marker for subtelomeric heterochromatin and chromosomal rearrangements in aphids of persicae group. *Chromosome Res.* **6**: 167–175.

Boehringer Mannheim (1992) *Nonradioactive* In Situ *Hybridization Application Manual*. Mannheim: Boehringer Mannheim.

Boehringer Mannheim (1996) *Nonradioactive* In Situ *Hybridization Application Manual*, 2nd Edn. Mannheim: Boehringer Mannheim.

Bracegirdle B, Bradbury S (1995) *Modern PhotoMICROgraphy*. Oxford: BIOS Scientific Publishers.

Bradbury S, Bracegirdle B (1998) *Introduction to Light Microscopy*. Oxford: BIOS Scientific Publishers.

Brandes A, Thompson H, Dean C, Heslop-Harrison JS (1997) Multiple repetitive DNA sequences in the paracentromeric regions of *Arabidopsis thaliana* L. *Chromosome Res.* **5**: 238–246.

Brandriff B, Gordon L, Trask B (1991) A new system for high-resolution DNA sequence mapping in interphase pronuclei. *Genomics* **10**: 75–82.

Brown TA (1999) *Genomes*. Oxford: Bios Scientific Publishers.

Camacho JPM, Carbrero J, Viseras E, Lópes-León MD, Navas-Castillo J, Alché JD (1991) G-banding in two species of grasshopper and its relationships to C, N and fluorescence banding techniques. *Genome* **34**: 638–643.

Carlemalm E, Villiger W (1989) Low temperature embedding. In: Bullock GR,

Petrutsz P, eds. *Techniques in Immunocytochemistry*. Vol 4. London: Academic Press, pp. 29–44.

Carter KC, Bowman D, Carrington W, Fogorty K, McNeil JA, Fay FS, Lawrence JB (1993) A three-dimensional view of precursor messenger RNA metabolism within the mammalian nucleus. *Science* **259**: 1330–1335.

Castilho A, Miller TE, Heslop-Harrison JS (1996) Physical mapping of translocation breakpoints in a set of wheat – *Aegilops umbellulata* – recombinant lines using *in situ* hybridization. *Theor. Appl. Genet.* **93**: 816–825.

Coen ES, Romero JM, Doyle S, Elliott R, Murphy G, Carpenter R (1990) *Floricaula:* a homeotic gene required for flower development in *Antirrhinum majus. Cell* **63**: 1311–1322.

Conger AD, Fairchild LM (1953) A quick freeze method for making smear slides permanent. *Stain Technol.* **28**: 281–283.

Cornish EC, Pettitt JM, Bonig I, Clarker AE (1987) Developmentally controlled expression of a gene associated with self incompatibility in *Nicotiana alata. Nature* **326**: 99–102.

Cox KH, DeLeon DV, Angerer LM, Angerer RC (1984) Detection of mRNAs in sea urchin embryos by *in situ* hybridization using asymmetric RNA probes. *Develop. Biol.* **101**: 485–503.

Cremer T, Lichter P, Borden J, Ward DC, Manuelidis L (1988) Detection of chromosome aberrations in metaphase and interphase tumor cells by *in situ* hybridization using chromosome-specific library probes. *Hum. Genet.* **80**: 235–246.

Cremer T, Kurz A, Zirbel R, Dietzel S, Rinke B, Schrock E, Speicher MR, Mathieu U, Jauch A, Emmerich P *et al*. (1993) Role of chromosome territories in the functional compartmentalization of the cell nucleus. *Cold Spring Harbor Symposia on Quantitative Biology LVIII*, pp. 777–791.

Cuadrado A, Schwarzacher T (1999) The chromosomal organization of simple sequence repeats in wheat and rye genomes. *Chromosoma* **107**: 587–594.

Darlington CD (1937) *Recent Advances in Cytology*. Churchill: London.

Davenloo P, Rosenberg AH, Dunn JJ, Studier FW (1984) Cloning and expression of the gene for bacteriophage T7 RNA polymerase. *Proc. Natl. Acad. Sci. USA* **81**: 2035–2039.

de Jong JH, van Eden J, Sybenga J (1989) Synaptonemal complex formation and metaphase I configuration patterns in a translocation heterozygote of rye (*Secale cereale*). *Genome* **32**: 72–81.

Dietzel S, Jauch A, Kienle D, Qu G, Holtgreve-Grez H, Eils R, Munkel C, Bittner M, Meltzer PS, Trent JM *et al*. (1998) Separate and variably shaped chromosome arm domains are disclosed by chromosome arm painting in human cell nuclei. *Chromosome Res.* **6**: 25–33.

Doležel J, Číhalíková J, Lucretti S (1992) A high yield procedure for isolation of metaphase chromosomes from root tips of *Vicia faba. Planta* **188**: 93–98.

Doležel J, Lucretti S, Schubert I (1994) Plant chromosome analysis and sorting by flow cytometry. *Critical Rev. Plant Sci.* **13**: 275–309.

Doudrick RL, Heslop-Harrison JS, Nelson CD, Schmidt T, Nance WL, Schwarzacher T (1995) Karyotype of Slash Pine (*Pinus elliottii* var. *elliottii*) using patterns of fluorescence *in situ* hybridization and fluorochrome banding. *J. Heredity* **86**: 289–296.

Dudin G, Cremer T, Schardin M, Hausmann M, Bier F, Cremer C (1987) A method for nucleic acid hybridisation to isolated chromosomes in suspension. *Hum. Genet.* **76**: 290–292.

Durrant I, Chadwick PME (1994) Hybridization of fluorescein-labeled oligonucleotide probes and enhanced chemiluminescence detection. In: Isaac PG, ed. *Protocols for Nucleic Acid Analysis by Nonradioactive Probes. Methods in Molecular Biology*, vol. 28. Totowa, NJ: Humana, pp. 141–148.

Feinberg AP, Vogelstein B (1983) A technique for radiolabeling DNA restriction endonuclease fragments to high specific activity. *Anal. Biochem.* **132**: 6–13. Addendum **137**: 266–267.

Ferguson-Smith MA (1991) Invited editorial: Putting the genetics back into cytogenetics. *Am. J. Hum. Genet.* **48**: 179–182.

Fidlerova H, Senger G, Kost M, Sanseau P, Sheer D (1994) Two simple procedures for releasing chromatin from routinely fixed cells for fluorescence *in situ* hybridization. *Cytogenet. Cell Genet.* **65**: 203–205.

Flavell AJ, Dunbar E, Anderson R, Pearce SR, Hartley R, Kumar A (1992) Ty1–copia group retrotransposons are ubiquitous and heterogeneous in higher plants. *Nucleic Acids Res.* **20**: 3639–3644.

Fobert PR, Coen ES, Murphy WP, Doonan JH (1994) Patterns of cell division revealed by transcriptional regulation of genes during the cell cycle in plants. *EMBO J.* **13**: 616–624.

Fobert PR, Gaudin V, Lunness P, Coen ES, Doonan JH (1996) Distinct classes of *cdc2*-related genes are differentially expressed during the cell division cycle in plants. *Plant Cell* **8**: 1465–1476.

Fransz PF, Alonso-Blanco C, Liharska TB, Peeters AJM, Zabel P, de Jong JH (1996) High-resolution physical mapping in *Arabidopsis thaliana* and tomato by fluorescence *in situ* hybridization to extended DNA fibres. *Plant J.* **9**: 421–430.

Fransz PF, de Jong JH, Zabel P (1998) Preparation of extended DNA fibres for high resolution mapping by fluorescence *in situ* hybridization (FISH). *Plant Molecular Biology Manual G5*, pp. 1–18.

Fuchs J, Houben A, Brandes A, Schubert I (1996) Chromosome 'painting' in plants – a feasible technique? *Chromosoma* **104**: 315–320.

Fuchs J, Kuhfittig S, Reuter G, Schubert I (1998) Chromosome painting in *Drosophila. Chromosome Res.* **6**: 335–336.

Fukui K, Nakayama S, eds (1996) *Plant Chromosomes: Laboratory Methods.* Boca Raton: CRC Press.

Galasso I, Schmidt T, Pignone D, Heslop-Harrison JS (1995) The molecular cytogenetics of *Vigna unguiculata* (L.) Walp: the physical organization and characterization of 18S-5.8S-25S rRNA genes, 5S rRNA genes, telomere-like sequences, and a family of centromeric repetitive DNA sequences. *Theor. Appl. Genet.* **91**: 928–935.

Gall JG, Pardue ML (1969) Formation and detection of RNA–DNA hybrid molecules in cytological preparations. *Proc. Natl. Acad. Sci. USA* **69**: 378–383.

Gatt M, Hammett K, Murray B (1999) Confirmation of ancient polyploidy in *Dahlia* (Asteraceae) species using genomic *in situ* hybridization. *Ann. Bot.* **84**: 39–48.

Geber G, Schweizer D (1988) Cytochemical heterochromatin differentiation in *Sinapis alba* (Cruciferae) using a simple air-drying technique for producing chromosome spreads. *Plant Syst. Evol.* **158**: 97–106.

Gerdes MG, Carter KC, Moen Jr PT, Lawrence JB (1994) Dynamic changes in the higher-level chromatin organization of specific sequences revealed by *in situ* hybridization to nuclear halos. *J. Cell Biol.* **126**: 289–304.

Giaid A, Gibson SJ, Steel J, Facer P, Polak JM (1990) The use of complementary RNA probes for the identification and localisation of peptide messenger RNA in the diffuse neuroendocrine system. *Soc. Exp. Biol. Seminar Ser.* **40**: 43–68.

Gillies CB (1981) Electron microscopy of spread maize pachytene synaptonemal complexes. *Chromosoma* **83**: 575–591.

Gomez MI, Islam-Faridi MN, Woo S-S, Schertz KF, Czeschin DJ, Zwick MS, Wing RA, Stelly DM, Price HJ (1997) FISH of a maize *sh2*-selected sorghum BAC on chromosomes of *Sorghum bicolor. Genome* **40**: 475–478.

Gosden JR (1997) *PRINS and* in situ *PCR Protocols.* New Jersey: Humana Press.

Gosden JR, Mitchell AR, Buckland RA, Clayton RP, Evans HJ (1975) The location of four human satellite DNAs on human chromosomes. *Exp. Cell Res.* **92**: 148–158.

Gosden J, Hanratty D, Starling J, Fantes J, Mitchell A, Porteous D (1991) Oligonucleotide-primed *in situ* DNA synthesis (PRINS): a method for chromosome mapping, banding, and investigation of sequence organization. *Cytogenet. Cell Genet.* **57**: 100–104.

Greider CW, Blackburn EH (1989) A telomeric sequence in the RNA of *Tetrahymena* telomerase required for telomere repeat synthesis. *Nature* **337**: 331–336.

Gurr E (1965) *The Rational Use of Dyes in Biology and General Staining Methods.* Baltimore: Williams and Wilkins.

Gyllenstain UB, Erlich HA (1988) Generation of single-stranded DNA by the polymerase chain reaction and its application to direct sequencing of the HLA-DQA locus. *Proc. Natl. Acad. Sci. USA* **85**: 7652–7656.

Haaf T, Ward DC (1994) Structural analysis of α-satellite DNA and centromere proteins using extended chromatin and chromosomes. *Hum. Mol. Gen.* **3**: 697–709.

Hadlaczky G, Bisztray G, Praznovsky T, Dudits D (1983) Mass isolation of plant chromosomes and nuclei. *Planta* **157**: 278–285.

Harper G, Hull R (1998) Cloning and sequence analysis of banana streak virus DNA. *Virus Genes* **17**: 271–278.

Harper G, Osuji JO, Heslop-Harrison JS, Hull R (1999) Integration of banana streak badnavirus into the *Musa* genome: molecular and cytogenetic evidence. *Virology* **255**: 207–213.

Harper ME, Saunders GF (1981) Localization of single copy DNA sequences on G-banded human chromosomes by *in situ* hybridization. *Chromosoma* **83**: 431–439.

Harris N, Mulcrone J, Grindley H (1990) Tissue preparation techniques for *in situ* hybridization studies of storage-protein gene expression during pea seed development. *Soc. Exp. Biol. Seminar Ser.* **40**: 175–188.

Harrison GE, Heslop-Harrison JS (1995) Centromeric repetitive DNA in the genus *Brassica*. *Theor. Appl. Genet.* **90**: 157–165.

Hawcroft DM (1996) *Electrophoresis: The Basics*. Oxford: BIOS Scientific Publishers.

Heiskanen M, Karhu R, Hellsten E, Peltonen L, Kallioniemi O-P, Palotie A (1994) High resolution mapping using fluorescent *in situ* hybridization to extended DNA fibers prepared from agarose-embedded cells. *BioTechniques* **5**: 928–933.

Heiskanen M, Kallioniemi O-P, Palotie A (1996) Fibre-FISH: experiences and a refined protocol. *Genetic Analysis* **12**: 179–184.

Hell SW, Stelzer EHK, Lindek S, Cremer C (1994) Confocal microscopy with an increased detection aperture: type-B 4Pi confocal microscopy. *Optics Letters* **19**: 222–224.

Heng HHQ, Squire J, Tsui L-C (1992) High resolution mapping of mammalian genes by *in situ* hybridization to free chromatin. *Proc. Natl. Acad. Sci. USA* **89**: 9509–9513.

Heng HHQ, Tsui L-C (1993) Modes of DAPI banding and simultaneous *in situ* hybridization. *Chromosoma* **102**: 325–332.

Herman B (1998) *Fluorescence Microscopy*, 2nd edn. Oxford: BIOS Scientific Publishers.

Herrington S, O'Leary J, eds. (1998) *PCR3. PCR In Situ Hybridization: A Practical Approach*. Oxford: Oxford University Press.

Herrmann RG, Martin R, Busch W, Wanner G, Hohmann U (1996) Physical and topographical mapping among Triticeae chromosomes. In: Heslop-Harrison JS, ed. *Unifying Plant Genomes: Comparisons, Conservation and Collinearity. 50th SEB Symposium*. SEB: London, pp. 25–30.

Heslop-Harrison JS, Schwarzacher T (1993) Molecular cytogenetics – biology and applications in plant breeding. *Chromosomes Today* **11**: 191–198.

Heslop-Harrison JS, Schwarzacher T (1996a) Genomic Southern and *in situ* hybridization for plant genome analysis. In: Jauhar PP, ed. *Methods of Genome Analysis in Plants*. Boca Raton: CRC, pp. 163–179.

Heslop-Harrison JS, Schwarzacher T (1996b) Flow cytometry and chromosome sorting. In: Fukui K, Nakayama S, eds. *Plant Chromosomes: Laboratory Methods*. Boca Raton: CRC, pp. 85–108.

Heslop-Harrison JS, Schwarzacher T, Leitch AR, Anamthawat-Jónsson K, Bennett MD (1988) A method of identifying DNA sequences in chromosomes of plants. European Patent Application Number 8828130.8.

Heslop-Harrison JS, Leitch AR, Schwarzacher T, Anamthawat-Jónsson K (1990) Detection and characterization of 1B/1R translocations in hexaploid wheat. *Heredity* **65**: 385–392.

Heslop-Harrison JS, Schwarzacher T, Anamthawat-Jónsson T, Leitch AR, Shi M, Leitch IJ (1991) *In situ* hybridization with automated chromosome denaturation. *Technique* **3**: 109–115.

Heslop-Harrison JS, Harrison GE, Leitch IJ (1992) Reprobing of DNA: DNA *in situ* hybridization preparations. *Trends Genet.* **8**: 372–373.

Heslop-Harrison JS, Brandes A, Taketa S, Schmidt T, Vershinin AV, Alkhimova EG, Kamm A, Doudrick RL, Schwarzacher T, Katsiotis A *et al.* (1997) The chromosomal distributions of Ty1–*copia* group retrotransposable elements in higher plants and their implications for genome evolution. *Genetica* **100**: 197–204.

Heslop-Harrison JS, Murata M, Ogura Y, Schwarzacher T, Motoyoshi F (1999) Polymorphisms and genomic organization of repetitive DNA from centromeric regions of *Arabidopsis thaliana* chromosomes. *Plant Cell* **11**: 31–42.

Heyderman E (1979) Immunoperoxidase technique in histopathology: applications, methods and controls. *J. Clin. Pathol.* **32**: 971–978.

Hopman AHN, Wiegant J, van Dujin A (1987) Mercurated nucleic acid probes, a new principle for non-radioactive *in situ* hybridization. *Exp. Cell Res.* **169**: 357–368.

Hopman AHN, Poddighe PJ, Smeets WAGB, Moesker O, Beck JLM, Vooijs GP, Ramaekers FCS (1989) Detection of numerical chromosome aberrations in bladder cancer by *in situ* hybridization. *Am. J. Path.* **135**: 1105–1117.

Hopman AHN, Poddighe P, Moesker O, Rameakers FCS (1992) Interphase cytogenetics: an approach to the detection of genetic aberrations in tumours. In: *Diagnostic Molecular Pathology: A Practical Approach.* Oxford: IRL, pp. 142–167.

Houseal TM, Klinger KW (1994) Commentary: What's in a spot? *Hum. Mol. Gen.* **3**: 1215–1216.

Houseal TW, Dackowski WR, Landes GM, Klinger KW (1994) High resolution mapping of overlapping cosmids by fluorescence *in situ* hybridization. *Cytometry* **15**: 193–198.

Hsu TC (1952) Mammalian chromosomes *in vitro*. 1. The karyotype of man. *J. Hered.* **43**: 167–172.

Ijdo JW, Wells RA, Baldini A, Reeders ST (1991) Improved telomere detection using a telomere repeat probe (TTAGGG)n generated by PCR. *Nucleic Acids Res.* **19**: 4780.

Isaac PG, ed (1994) *Protocols for Nucleic Acid by Nonradioactive Probes. Methods in Molecular Biology 28.* Totoyo, New Jersey: Humana Press.

Jackson DP (1992) *In situ* hybridisation in plants. In: Bowles DJ, Gurr SJ, McPherson M, eds. *Molecular Plant Pathology: A Practical Approach.* Oxford: Oxford University Press, pp. 163–174.

Jauch A, Daumer C, Lichter P, Murken J, Schroeder KT, Cremer T (1990) Chromosomal *in situ* suppression hybridization of human gonosomes and autosomes and its use in clinical cytogenetics. *Hum.Genet.* **85**: 145–150.

Jiang J, Gill BS (1993) Sequential chromosome banding and *in situ* hybridization analysis. *Genome* **36**: 792–795.

Jiang J, Gill BS, Wang G-L, Ronald PC, Ward DC (1995) Metaphase and interphase fluorescence *in situ* hybridization mapping of the rice genome with bacterial artificial chromosomes. *Proc. Natl. Acad. Sci. USA* **92**: 4487–4491.

John HA, Birnstiel ML, Jones KW (1969) RNA–DNA hybrids at the cytological level. *Nature* **223**: 582–587.

Jones GH, Croft JA (1986) Surface spreading of synaptonemal complexes in locusts. II. Zygotene pairing behaviour. *Chromosoma* **93**: 489–495.

Joos S, Scherthan H, Speicher MR, Schlegel J, Cremer T, Lichter P (1993) Detection of amplified genome sequences by reverse chromosome painting using genomic tumor DNA as probe. *Hum. Genetics* **90**: 584–589.

Kallioniemi A, Kallioniemi O-P, Sudar D, Rutovitz D, Gray JW, Waldman F, Pinkel D (1992) Comparative genomic hybridization for molecular cytogenetic analysis of solid tumors. *Science* **258**: 818–821.

Kamm A, Galasso I, Schmidt T, Heslop-Harrison JS (1995) Analysis of a repetitive DNA family from *Arabidopsis arenosa* and relationships between *Arabidopsis* species. *Plant Mol. Biol.* **27**: 853–862.

Kipling D, Wilson HE, Mitchell AR, Taylor BA, Cooke HJ (1994) Mouse centromere mapping using oligonucleotide probes that detect variants of the minor satellite. *Chromosoma* **103**: 46–55.

Koch J, Mogensen J, Pedersen S, Fischer H, Hindkj'r J, Kolvraa S, Bolund L

(1992). Fast one-step procedure for the detection of nucleic acids *in situ* by primer-induced sequence-specific labeling with fluorescein-12–dUTP. *Cytogenet. Cell Genet.* **60**: 1–3.

Kosina R, Heslop-Harrison JS (1996) Molecular cytogenetics of an amphiploid trigeneric hybrid between Triticum durum, Thinopyrum distichum and Lophopyrum elongatum. *Ann. Bot.* **78**: 583–589.

Kubis S, Schmidt T, Heslop-Harrison JS (1998) Repetitive DNA elements as a major component of plant genomes. *Ann. Bot.* **82** (Suppl.): 45–55.

Kurz A, Lampel S, Nickolenko JE, Bradl J, Benner A, Zirbel RM, Vremer T, Lichter P (1996) Active and inactive genes localize preferentially in the periphery of chromosome territories. *J. Cell Biol.* **135**: 1195–1205.

Landegent JE, Jansen in de Wal N, Baan RA, Hoeijmakers JHJ, Ploeg M (1984) 2-Acetylaminofluorene-modified probes for the indirect hybridocytochemical detection of specific nucleic acid sequences. *Experimental Cell Res.* **153**: 61–72.

Langer PR, Waldrop AK, Ward DC (1981) Enzymatic synthesis of biotin labeled polynucleotides: novel nucleic acid affinity probes. *Proc. Natl. Acad. Sci. USA* **78**: 6633–6637.

Langer-Safer PR, Levine M, Ward DC (1982) Immunocytological method for mapping genes on *Drosophila* polytene chromosomes. *Proc. Natl. Acad. Sci. USA.* **79**: 4381–4385.

Lawrence JB, Singer RH, McNeil JA (1990) Interphase and metaphase resolution of different distances within the human distrophin gene. *Science* **249**: 928–932.

Le Beau MM, Espinosa III R, Neuman WL, Stock W, Roulston D, Larson RA, Keinanen M, Westbrook CA (1993) Cytogenetic and molecular dilineation of the smallest commonly deleted region of chromosome 5 in malignant myeloid diseases. *Proc. Natl Acad. Sci. USA* **90**: 5484–5488.

Lehfer H, Wanner G, Herrmann RG (1991) Physical mapping of DNA sequences on plant chromosomes by light microscopy and high resolution scanning electron microscopy. *Plant Mol. Biol.* **2**: 277–284.

Leitch AR, Mosgöller W, Schwarzacher T, Bennett MD, Heslop-Harrison JS (1990) Genomic *in situ* hybridization to sectioned nuclei shows chromosome domains in grass hybrids. *J. Cell Sci.* **95**: 335–341.

Leitch AR, Schwarzacher T, Mosgöller W, Bennett MD, Heslop-Harrison JS (1991) Parental genomes are separated throughout the cell cycle in a plant hybrid. *Chromosoma* **101**: 206–213.

Leitch AR, Schwarzacher T, Wang ML, Leitch IJ, Surlan-Momirovic G, Moore G, Heslop-Harrison JS (1993) Molecular cytogenetic analysis of repeated sequences in a long term wheat suspension culture. *Plant Cell, Tiss. Org. Cult.* **33**: 287–296.

Leitch AR, Schwarzacher T, Jackson D, Leitch IJ (1994a) In Situ *Hybridization.* Oxford: BIOS Scientific Publishers.

Leitch AR, Brown JKM, Mosgöller W, Schwarzacher T, Heslop-Harrison JS (1994b) The spatial localization of homologous chromosomes in human fibroblasts at mitosis. *Hum. Genet.* **93**: 275–280.

Leitch IJ, Heslop-Harrison JS (1992) Physical mapping of the 18S-5.8S-26S rRNA genes in barley by *in situ* hybridization. *Genome* **35**: 1013–1018.

Lengauer C, Eckelt A, Weith A, Endlich N, Ponelies N, Lichter P, Greulich KO, Cremer T (1991) Painting of defined chromosomal regions by *in situ* suppression hybridization of libraries from laser-microdissected chromosomes. *Cytogenet. Cell Genet.* **56**: 27–30.

Lengauer C, Green ED, Cremer T (1992) Fluorescence *in situ* hybridization of YAC clones after Alu-PCR amplification. *Genomics* **13**: 826–828.

Lichter P (1997) Multicolour FISHing: what's the catch? *Trends Genet.* **13**: 475–479.

Lichter P, Cremer T, Borden J, Manuelidis L, Ward DC (1988) Delineation of individual human chromosomes in metaphase and interphase cells by *in situ* suppression hybridization using recombinant DNA libraries. *Hum. Genet.* **80**: 224–234.

Lichter P, Tang C-JC, Call K, Hermanson G, Evans GA, Housman D, Ward DC (1990) High-resolution mapping of human chromosome 11 by *in situ* hybridization with cosmid clones. *Science* **247**: 64–69.

Lima-Brito JE, Guedes-Pinto H, Harrison GE, Heslop-Harrison JS (1997) Molecular cytogenetic analysis of durum wheat × tritordeum hybrid. *Genome* **40**: 362–369.

Liu YG, Whittier RF (1994) Rapid preparation of megabase plant DNA from nuclei in agarose plugs and microbeads. *Nucleic Acids Res.* **22**: 2168–2169.

Liyanage M, Coleman A, du Manoir S, Veldman T, McCormack S, Dickson RB, Barlow C, Wynshaw-Boris A, Janz S, Wienberg J, *et al.* (1996) Multicolour spectral karyotyping of mouse chromosomes. *Nature Genetics* **14**: 312–315.

Lopez-Leon MD, Neves N, Schwarzacher T, Heslop-Harrison JS, Hewitt GM, Camacho JPM (1994) Possible origin of a B chromosome deduced from its DNA composition using double FISH technique. *Chromosome Res.* **2**: 87–92.

Macas J, Doležel J, Gualberti G, Pich U, Schubert I, Lucretti S (1995) Primer-induced labeling of pea and field bean chromosomes *in situ* and in suspension. *BioTechniques* **19**: 402–408.

McFadden GI (1989) *In situ* hybridization in plants: from macroscopic to ultrastructural resolution. *Cell Biol. Int. Rep.* **13**: 3–21.

McFadden GI, Bonig I, Clarke AE (1990) Double label *in situ* hybridization for electron microscopy. *Trans. Roy. Microscopical Soc.* **1**: 683–688.

Macgregor HC, Varley JM (1988) *Working with Animal Chromosomes.* 2nd Edn. Chichester: Wiley.

McPherson MJ, Möller S (2000) *PCR: The Basics.* Oxford: BIOS Scientific Publishers.

Maluszynska J, Heslop-Harrison JS (1991) Localization of tandemly repeated DNA sequences in *Arabidopsis thaliana. Plant J.* **1**: 159–166.

Maniatis TA, Jeffrey A, Kleid DG (1975) Nucleotide sequence of the rightward operator of phage λ. *Proc. Natl. Acad. Sci. USA* **72**: 1184–1188.

Martin R (1996) *Gel Electrophoresis: Nucleic Acids.* Oxford: BIOS Scientific Publishers

Meinkoth J, Wahl G (1984) Hybridization of nucleic acids immobilized on solid supports. *Anal. Biochem.* **138**: 267–284.

Moens PB, Pearlman RE (1989) Satellite DNA I in chromatin loops of rat pachytene chromosomes and in spermatids. *Chromosoma* **98**: 287–294.

Moses MJ (1977) Synaptonemal complex karyotyping of spermatocytes of the Chinese hamster (*Cricetulus griseus*). *Chromosoma* **60**: 99–125.

Mouras A, Saul MW, Essad S, Potrykus I (1987) Localization by *in situ* hybridization of a low copy chimaeric resistance gene introduced into plants by direct gene transfer. *Mol. Gen. Genet.* **207**: 204–209.

Mukai Y, Nakahara Y, Yamamoto M (1993) Simultaneous discrimination of the three genomes in hexaploid wheat by multicolour fluorescence *in situ* hybridization using total genomic and highly repeated DNA probes. *Genome* **36**: 489–494.

Müller S, Rocchi M, Ferguson-Smith MA, Wienberg J (1997) Towards a multicolor chromosome bar code for the entire human karyotype by fluorescence *in situ* hybridization. *Hum. Genet.* **100**: 271–278.

Murata M (1983) Staining air dried protoplasts for study of plant chromosomes. *Stain Technology* **58**: 101–106.

Nanda I, Zischler H, Epplen C, Guttenbach M, Schmid M (1991) Chromosomal organization of simple repeated DNA sequences used for DNA fingerprinting. *Electrophoresis* **12**: 193–203.

Nederlof PM, van der Flier S, Wiegant J, Raap AK, Tanke HJ, Ploem JS, van der Ploeg M (1990) Multiple fluorescence *in situ* hybridization. *Cytometry* **11**: 126–131.

Neves N, Viegas W, Heslop-Harrison JS (1995a) Organization of chromatin at interphase in wheat × rye hybrids and in triticale. *Chromosome Res.* **3** (Suppl 1): 91.

Neves N, Heslop-Harrison JS, Viegas W (1995b) rRNA gene activity and control of expression mediated by methylation and imprinting during embryo development in wheat × rye hybrids. *Theor. Appl. Genet.* **91**: 529–533.

Neves N, Castilho A, Silva M, Heslop-Harrison JS, Viegas W (1996) Genomic interactions: gene expression, DNA methylation and nuclear architecture. *Chromosomes Today* **12**: 182–200.

Newton CR, Graham A (1997) *PCR*, 2nd edn. Oxford: BIOS Scientific Publishers.

Old RW, Primrose SB (1992) *Principles of Gene Manipulation*. 2nd Edn. Oxford: Blackwells.

Orgaard M, Jacobsen N, Heslop-Harrison JS (1995) The hybrid origin of two cultivars of *Crocus* (Iridaceae) analysed by molecular cytogenetics including genomic Southern and *in situ* hybridization. *Ann. Bot.* **76**: 253–262.

Osuji JO, Harrison G, Crouch J, Heslop-Harrison JS (1997) Identification of the genomic constitution of Musa L. lines (bananas, plantains and hybrids) using molecular cytogenetics. *Ann. Bot.* **80**: 787–793.

Parra I, Windle B (1993) High resolution visual mapping of stretched DNA by fluorescent hybridization. *Nature Genetics* **5**: 17–21.

Pearse AGE (1968) *Histochemistry: Theoretical and Applied*. 3rd Edn. Baltimore: Williams and Watkins.

Pearse AGE (1972) *Histochemistry: Theoretical and Applied*. 4th Edn. London: Churchill Livingstone.

Pedersen C, Rasmussen SK, Linde-Laursen I (1996) Genome and chromosome identification in cultivated barley and related species of the Triticeae (Poaceae) by *in situ* hybridization with the GAA-satellite sequence. *Genome* **39**: 93–104.

Pich U, Meister A, Macas J, Dolezel J, Lucretti S, Schubert I (1995) Primed *in situ* labelling facilitates flow sorting of similar sized chromosomes. *Plant J.* **7**: 1039–1044.

Pinkel D, Straume T, Gray JW (1986) Cytogenetic analysis using quantitative, high-sensitivity, fluorescence hybridization. *Proc. Natl. Acad. Sci. USA* **83**: 2934–2938.

Pinkel D, Landegent J, Collins C, Fuscoe J, Segraves R, Lucas J, Gray JW (1988) Fluorescence *in situ* hybridization with human chromosome-specific libraries: detection of trisomy 21 and translocations of chromosome 4. *Proc. Natl. Acad. Sci. USA* **85**: 9138–9142.

Ponder BA, Wilkinson MM (1981) Inhibition of endogenous tissue alkaline phosphatase with the use of alkaline phosphatase conjugates in immunohisto-chemistry. *J. Histochem.* **29**: 981–984.

Popp S, Scholl IIP, Loos P, Jauch A, Stelzer E, Cremer C, Cremer T (1990) Distribution of chromosome 18 and X centric heterochromatin in the inter-phase nucleus of cultured human cells. *Exp. Cell Res.* **189**: 1–12.

Rawlins DJ, Highett MI, Shaw P (1991) Localization of telomeres in plant inter-phase nuclei by *in situ* hybridization and 3D confocal microscopy. *Chromosoma* **100**: 424–431.

Reader SM, Abbo S, Purdie KA, King IP, Miller TE (1994) Direct labelling of plant chromosomes by rapid *in situ* hybridization. *Trends Genet.* **10**: 265–266.

Ribeiro-Carvalho C, Guedes-Pinto H, Harrison GE, Heslop-Harrison JS (1997) Wheat–rye chromosome translocations involving small terminal and intercalary rye chromosome segments in the Portuguese wheat landrace Barbela. *Heredity* **78**: 539–546.

Richards EJ, Vongs A, Walsh M, Yang J, Chao S (1993) Substructure of telomere repeat arrays. In: Heslop-Harrison JS, Flavell RB, eds. *The Chromosome*. Oxford: BIOS Scientific Publishers, pp. 103–114.

Rosen B, Beddington RSP (1993) Whole-mount *in situ* hybridization in the mouse embryo: gene expression in three dimensions. *Trends Genet.* **9**: 162.

Rozen S, Skaletsky HJ (1998) Primer 3. Code available at http://www.genome.wi.mit.edu/genome_software/other/primer3.html

Ruzin SE (1999) *Plant microtechnique and microscopy*. New York: Oxford University Press.

Sambrook J, Fritsch EF, Maniatis T (1989) *Molecular Cloning: a Laboratory Manual*. Cold Spring Harbor, New York: Cold Spring Harbor Laboratory Press.

Schardin M, Cremer T, Hager HD, Lang M (1985) Specific staining of human chromosomes in Chinese hamster × man hybrid cell lines demonstrates inter-phase chromosome territories. *Hum. Genet.* **71**: 281–287.

Scherthan H, Cremer T, Arnason U, Weier H-U, Lima-de-Faria A, Fronicke L

(1994) Comparative chromosome painting discloses homologous segments in distantly related mammals. *Nature Genetics* **6**: 342–347.

Schmidt T, Heslop-Harrison JS (1996a) The physical and genomic organization of microsatellites in sugar beet. *Proc. Natl. Acad. Sci. USA* **93**: 8761–8765.

Schmidt T, Heslop-Harrison JS (1996b) High resolution mapping of repetitive DNA by *in situ* hybridization: molecular and chromosomal features of prominent dispersed and discretely localized DNA families from the wild beet species *Beta procumbens*. *Plant Mol. Biol.* **30**: 1099–1114.

Schmidt T, Heslop-Harrison, JS (1998) Genomes, genes and junk: the large scale organization of plant chromosomes. *Trends Plant Sci.* **3**: 195–199.

Schmidt T, Schwarzacher T, Heslop-Harrison JS (1994) Physical mapping of rRNA genes by fluorescent *in situ* hybridization and structural analysis of 5S rRNA genes and intergenic spacer sequences in sugar beet (*Beta vulgaris*). *Theor. Appl. Genet.* **88**: 629–636.

Schmidt T, Kubis S, Heslop-Harrison JS (1995) Analysis and chromosomal localization of retrotransposons in sugar beet (*Beta vulgaris* L.): LINEs and Ty1-copia-like elements as major components of the genome. *Chromosome Res.* **3**: 335–345.

Schmitz GG, Waiter T, Seibl R, Kessler C (1991) Nonradioactive labeling of oligonucleotides *in vitro* with the hapten digoxigenin by tailing with terminal transferase. *Anal. Biochem.* **192**: 222–231.

Schröck E, du Manoir S, Veldman T, Schoell B, Wienberg J, Ferguson-Smith MA, Ning Y, Ledbetter DH, Bar-Am I, Soenksen D, *et al.* (1996) Multicolor spectral karyotyping of human chromosomes. *Science* **273**: 494–497.

Schwarzacher HG (1974) Preparation of metaphase chromosomes. In: Schwarzacher HG, Wolf U, eds. *Methods in Human Cytogenetics*. Berlin: Springer, pp. 71–81.

Schwarzacher HG, Wolf U, eds (1974) *Methods in Human Cytogenetics*. Springer: Berlin.

Schwarzacher T (1996) The physical organization of Triticeae chromosomes. In: Heslop-Harrison JS, ed. *Unifying Plant Genomes. 50th SEB Symposium Unifying Plant Genomes*. Cambridge: Company of Biologists, pp. 71–75.

Schwarzacher T (1997) Three stages of meiotic homologous chromosome pairing in wheat: cognition, alignment and synapsis. *Sex. Plant Reprod.* **10**: 324–331.

Schwarzacher T, Heslop-Harrison JS (1991) *In situ* hybridization to plant telomeres using synthetic oligomers. *Genome* **34**: 317–323.

Schwarzacher T, Heslop-Harrison JS (1994) Direct fluorochrome labelled DNA probes for direct fluorescent *in situ* hybridization to chromosomes. In: Isaac PG, ed. *Protocols for Nucleic Acid Analysis by Nonradioactive Probes. Methods in Molecular Biology*. Vol 28. Totowa, New Jersey: Humana, pp. 167–176.

Schwarzacher T, Ambros P, Schweizer D (1980) Application of Giemsa banding to orchid karyotype analysis. *Plant Syst. Evol.* **134**: 293–297.

Schwarzacher T, Cram LS, Meyne J, Moyzis RK (1988) Characterization of human heterochromatin by *in situ* hybridization with satellite DNA clones. *Cytogenet. Cell Genet.* **47**: 192–196.

Schwarzacher T, Leitch AR, Bennett MD, Heslop-Harrison JS (1989) *In situ* localization of parental genomes in a wide hybrid. *Ann. Bot* **64**: 315–324.

Schwarzacher T, Anamthawat-Jónsson K, Harrison GE, Islam AKMR, Jia JZ, King IP, Leitch AR, Miller TE, Reader SM, Rogers WJ, *et al.* (1992a) Genomic *in situ* hybridization to identify alien chromosomes and chromosome segments in wheat. *Theor. Appl. Genet.* **84**: 778–786.

Schwarzacher T, Heslop-Harrison JS, Anamthawat-Jónsson K, Finch RA, Bennett MD (1992b) Parental genome separation in reconstructions of somatic and premeiotic metaphases of *Hordeum vulgare* × *H. bulbosum*. *J. Cell Sci.* **101**: 13–24.

Schwarzacher T, Leitch AR, Heslop-Harrison JS (1994) DNA:DNA *in situ* hybridization – methods for light microscopy. In: Harris N, Oparka KJ. eds. *Plant Cell Biology: A Practical Approach*. Oxford: Oxford University Press, pp. 127–155.

Schwarzacher T, Wang ML, Leitch AR, Miller N, Moore G, Heslop-Harrison JS

(1997) Flow cytometric analysis of the chromosomes and stability of a wheat cell-culture line. *Theor. Appl. Genet.* **94**: 91–97.

Sjöberg A, Peelman LJ, Chowdhary BP (1997) Application of three different methods to analyse fibre-FISH results obtained using four lambda clones from the porcine MHCIII region. *Chromosome Res.* **5**: 247–253.

Speed RM (1982) Meiosis in the foetal mouse ovary. I. An analysis at the light microscope level using surface-spreading. *Chromosoma* **85**: 427–437.

Speicher MR, Ballard SG, Ward DC (1996) Computer image analysis of combinatorial multi-fluor FISH. *Bioimaging* **4**: 52–64.

Straus W (1971) Inhibition of peroxidase by methanol and by methanol-nitroferricyanide for use in immunoperoxidase procedures. *J. Histochem. Cytochem.* **19**: 682–688.

Straus W (1972) Phenylhydrazine, an inhibitor of horseradish peroxidase for use in immunoperoxidase procedures. *J. Histochem. Cytochem.* **20**: 949–951.

Strehl S, Ambros PF (1993) Fluorescent *in situ* hybridization combined with immunohistochemistry for highly sensitive detection of chromosome 1 aberrations in neuroblastoma. *Cytogenet. Cell Genet.* **63**: 24–28.

Tautz D, Pfeiffle C (1989) A non-radioactive *in situ* hybridization method for the localization of specific RNAs in *Drosophila* embryos reveals translation control of the segmentation gene *hunchback*. *Chromosoma* **98**: 81–85.

Tautz D, Hulskamp M, Sommer RJ (1992) Whole mount *in situ* hybridization in *Drosophila*. In: Wilkinson DG, ed. In Situ *Hybridization: A Practical Approach*. Oxford: IRL Press, pp. 61–73.

Telenius H, Palmear AH, Tunnacliffe A, Carter NP, Behmel A, Ferguson-Smith MA, Nordenskjold M, Pfragner R, Ponder BAJ (1992) Cytogenetic analysis by chromosome painting using DOP-PCR amplified flow-sorted chromosomes. *Genes Chromosomes Cancer* **4**: 257–263.

Tu CPD, Cohen SN (1980) 3′-end labeling of DNA with [-32P] cordy-cepin-5′-triphosphate. *Gene* **10**: 177.

Vass K, Berger M, Nowak TS Jr, Welch WJ, Lassmann H (1989) Induction of stress protein HSP 70 nerve cells after status epilepticus in rat. *Neurosci. Lett.* **100**: 259–264.

Vaughan HF, Heslop-Harrison JS, Hewitt GM (1999) The localisation of mitochondrial sequences to chromosomal DNA in Orthopterans. *Genome* (in press).

Verma RS, Babu A (1995) *Human Chromosomes: Principles and Techniques*. New York: McGraw Hill.

Vershinin AV, Schwarzacher T, Heslop-Harrison JS (1995) The large scale genomic organization of repetitive DNA families at the telomeres of rye chromosomes. *Plant Cell* **7**: 1823–1833.

Viegas-Pequignot E (1992) *In situ* hybridization to chromosomes with biotinylated probes. In: Wilkinson DG, ed. In situ *Hybridization: A Practical Approach*. Oxford: IRL, pp. 137–158.

Vischer NOE, Huls PG, Woldringh CL (1994) Object-Image: an interactive image analysis program using structured point collection. *Binary* **6**: 35–41.

Visser AE, Eils R, Jauch A, Little G, Bakker PJM, Cremer T, Aten JA (1998) Spatial distributions of early and late replicating chromatin in interphase chromosome territories. *Exp. Cell Res.* **243**: 398–407.

Wachtler F, Mosgöller W, Schwarzacher HG (1990) Electron microscopic *in situ* hybridization and autoradiography: localization and transcription of rDNA in human lymphocyte nucleoli. *Exp. Cell Res.* **187**: 346–348.

Wachtler F, Schöfer C, Mosgöller W, Weipoltshammer K, Schwarzacher HG, Guichaoua M, Hartung M, Stahl A, Berge-Lefranc JL, Gonzalez I, *et al*. (1992) Human ribosomal RNA gene repeats are localized in the dense fibrillar component of nucleoli: light and electron microscopic *in situ* hybridization in human sertoli cells. *Exp. Cell Res.* **198**: 135–143.

Wallace RB, Johnson MJ, Hirose T, Miyake T, Kawashima EH, Itakura K (1981) The use of synthetic oligonucleotides as hybridization probes. II. Hybridization of oligonucleotides of mixed sequence to rabbit α-globin DNA. *Nucleic Acids Res.* **9**: 879–894.

Wang M, Duell T, Gray JW, Weier H-UG (1996) High sensitivity, high resolution physical mapping by fluorescence *in situ* hybridization on to individual straightened DNA molecules: Bioimaging DNA molecules. *Bioimaging* **4**: 73–83.

Wang ML, Leitch AR, Schwarzacher T, Heslop-Harrison JS, Moore G (1992) Construction of a chromosome-enriched *Hpa*II library from flow-sorted wheat chromosomes. *Nucleic Acids Res.* **20**: 1897–1901.

Warford A, Lauder I (1991) *In situ* hybridization in perspective. *J. Clin. Pathol.* **44**: 177–181.

Wetmer JG, Ruyechen WT, Douthart RJ (1981) Denaturation and renaturation of *Penicillium chrysogenum* mycophage double-stranded ribonucleic acid in tetra alkylammonium salt solutions. *Biochemistry* **20**: 2999–3002.

Wienberg J, Jauch A, Ludecke H-J, Senger G, Horsthemke B, Claussen U, Cremer T, Arnold N, Lengauer C (1994) The origin of human chromosome 2 analyzed by comparative chromosome mapping with a DNA microlibrary. *Chromosome Res.* **2**: 405–410.

Wilkinson DG, ed (1992) In situ *Hybridization: A Practical Approach*. Oxford: IRL Press.

Wilson WD, Tanious FA, Barton HJ, Jones RL, Fox K, Wydra RL, Strekowski L (1990) DNA sequence dependent binding mode of 4',6-diamidino-2-phenylindole (DAPI). *Biochemistry* **29**: 8452–8486.

Wood GS, Warnke R (1981) Suppression of endogenous avidin binding in tissues and its relevance to biotin-avidin detection systems. *J. Histochem. Cytochem.* **29**: 1196.

Yang F-T (1998) *Chromosome Evolution of the Muntjacs: Inferences from Molecular Cytogenetics.* PhD Thesis, Department of Pathology, University of Cambridge.

Buffers and other solutions

For making buffers and solutions use highly purified water, typically double distilled or purified by reverse osmosis and other methods. The quality of tap water can vary: water companies may alter chemical treatments seasonally. Not all chemicals are efficiently removed by purification treatments. To inhibit growth of microorganisms, we usually use sterile, purified water to make up solutions. Simple salt and Tris solutions should be autoclaved before storage. Information about biological buffers is found in Beynon and Easterby (1996).

1 M Tris-HCl buffer

Many molecular biology protocols use buffers based on Tris (tris-[hydroxymethyl]amino-methane). This has good buffering capacity in the biologically relevant pH range, and it does not change hybridization stringency nor inhibit enzymes, because it does not contain the elemental cations found in many buffers. The pH of Tris solutions changes widely with temperature, decreasing by 0.31 units with a 10°C increase in temperature; it also decreases by 0.4 units with dilution from 1 M to 10 mM. The pH given in protocols in this book refers to the undiluted stock solution (1 M for Tris) at 20°C, with ±0.1 unit accuracy (i.e., the pH is not that of the solution after dilution).

Tris-HCl ('Tris hydrochloride') buffer stock solutions (normally 1 M) can be:

(1) Purchased as sets (e.g. from Sigma);
(2) Made by mixing stock solutions of 1 M Tris base (121.1 g l^{-1}) and 1 M Tris-HCl (157.6 g l^{-1}) (see *Table A1*, based on data from Sigma);
(3) Made by mixing 1 M Tris base with 1 M HCl. To make a 1 M solution, dissolve Tris base (121.1 g) in 900 ml water, add 65 ml (pH 7.4), 55 ml (pH 7.6) or 35 ml (pH 8.0) 1 M HCl, adjust pH with an additional 5–8 ml of 1 M HCl, and make up to 1 l with water.

Some electrodes do not measure the pH of Tris solutions accurately, and electrodes specified for Tris should be used. Solutions made by mixing Tris base and Tris-HCl will be accurate enough to use with pH measurement. Tris-HCl buffer should be sterilized by autoclaving.

500 mM EDTA pH 8

Dissolve 186.1 g disodium ethylene diamine tetraacetate·2H$_2$O in 800 ml of water. Add about 10 g NaOH pellets, stir on a magnetic stirrer and adjust to pH 8 with further NaOH (about 10 g pellets). Make up to 1 l and sterilize by autoclaving.

10× TE-buffer (Tris-EDTA buffer)

Mix 100 mM Tris-HCl buffer of the required pH with 10 mM EDTA pH 8.

Table A1 Tris buffers made by mixing Tris-hydrochloride (Tris-HCl) with Tris base.

To prepare 1 l of 1 M Tris-HCl buffer at pH	1 M Tris-HCl (ml)	1 M Tris base (ml)
7.0	948.8	51.2
7.2	917.4	82.6
7.4	869.6	130.4
7.6	806.4	193.6
7.8	719.9	280.1
8.0	612.4	387.6

10× NTE-buffer (NaCl-Tris-EDTA buffer)

Also called STE-buffer (sodium-Tris-EDTA buffer).

Mix 5 M NaCl, 100 mM Tris-HCl pH 7.5, 10 mM EDTA pH 8.

20× SSC (saline sodium citrate or standard saline citrate)

This is a widely used, weak buffer which is used for many washes and to control stringency during *in situ* hybridization. The 20× stock solution consists of 3 M sodium chloride and 300 mM trisodium citrate. 20× SSC is used for two reasons: it is the strongest that readily goes into solution, while fungi and bacteria do not grow readily in this strength.

To make the stock, dissolve 175.3 g NaCl and 88.2 g $Na_3C_6H_5O_7 \cdot 2\ H_2O$ in 900 ml water. Adjust to pH 7 with NaOH or HCl if necessary, make up to 1 l and sterilize by autoclaving.

Phosphate buffers

Phosphate buffers (normally sodium phosphate, although potassium or mixtures are used regularly) are made by mixing disodium phosphate and monopotassium phosphate or monosodium phosphate of the required molarity in water to reach the required pH. Phosphate buffer can be kept for a short time at 4°C.

For 1 M stock solutions:

Dissolve 14.2 g Na_2HPO_4 (or 26.8 g $Na_2HPO_4 \cdot 7H_2O$) in 100 ml water (final volume).
Dissolve 13.6 g KH_2PO_4 in 100 ml water (final volume).
Dissolve 12.0 g NaH_2PO_4 (or 13.8 g $NaH_2PO_4 \cdot H_2O$) in 100 ml water (final volume).

To make the working buffer for the:

100 mM phosphate buffer at pH 6.9: mix 60 ml of 100 mM Na_2HPO_4 and 40 ml of 100 mM KH_2PO_4.

75 mM sodium phosphate at pH 7: Mix 61.5 ml of 75 mM Na_2HPO_4 and 38.5 ml of 75 mM NaH_2PO_4.

6 mM sodium phosphate buffer at pH 7.4: mix 81 ml of 60 mM Na_2HPO_4 and 19 ml of 60 mM NaH_2PO_4, dilute 1:10 with water.

10× PBS (phosphate-buffered saline)

PBS is widely used in cell culture and for washing or treatments of preparations, and contains 120 mM saline in phosphate buffer. Stock solutions are usually made at 10× and diluted to 1× before use. PBS can be bought ready mixed as a concentrate or powder for dilution (e.g. from Sigma). Alternatively, make a $10\times$ stock solution, autoclave and keep at 4°C and dilute before use. Many formulations include 2.7 mM KCl, but this is not necessary in the applications in this book; the NaCl concentration can be 138 mM or 145 mM in other formulations.

To make $10\times$ PBS (1.3 M NaCl, 70 mM Na_2HPO_4, 30mM NaH_2PO_4), add 76.0 g NaCl, 12.46 g $Na_2HPO_4\cdot2\ H_2O$ and 4.68 g $NaH_2PO_4\cdot2\ H_2O$ to 900 ml water, adjust pH to 7.4 with 1 M NaOH or HCl, and make up to 1 l with water before autoclaving.

$1\times$ TBS (Tris buffered saline)

TBS is recommended as an alternative to PBS or SSC in some applications. Make up 50 mM Tris-HCl with 150 mM NaCl and adjust pH to 7.6 with hydrochloric acid.

$1\times$ Detection buffer

Detection buffer is used for rinsing slides after detection, with a detergent to help uniform spreading and rinsing of previous solutions, and a high salt concentration to avoid removing labels. It consists of $4\times$ SSC (diluted from $20\times$ SSC) with 0.2% (v/v) Tween 20, but PBS or other buffers can also be used.

$10\times$ Enzyme buffer pH 4.6.

(1) Make solution A (100 mM citric acid) by dissolving 2.1 g $C_6H_8O_7\cdot H_2O$ in 100 ml of water (final volume).

(2) Make solution B (100 mM tri-sodium citrate) by dissolving 2.94 g $C_6H_5O_7Na_3$ $\cdot2H_2O$ in 100 ml (final volume).

(3) Mix 40 ml of solution A and 60 ml of solution B and dilute 1:10 with water for use.

Names and addresses of major suppliers

We have listed selected major suppliers of molecular and molecular cytogenetic reagents, kits, instrumentation and services; we have tried to give addresses in the US, UK and Japan, as well as headquarters, but not telephone numbers which change rapidly. Many products are available from distributors, particularly outside their country of origin. In this age of flux, companies move, merge, split, vanish or change names regularly. New companies may supply the most innovative products. We recommend using the Internet (see Chapter 13) routinely to find company details and contacts, local distributors or suppliers of unusual reagents and equipment. The site http://www.chromosome.net (coordinated by Dr Shigeki Nakayama, and used in the compilation of this list) carries an up-to-date list of supplier net addresses, while www.kaker.com has an extensive list of microscopy suppliers. For a few reagents (e.g. human chromosome paints), patents restrict availability in some countries.

Adobe Systems: Software including Adobe Photoshop and Adobe Illustrator.

1585 Charleston Road, PO Box 7900, Mountain View, CA 94039–7900, USA.
Adobe House, Mid New Cultins, Edinburgh, EH11 4DU, UK.
PO Box 20, 300AA Rotterdam, the Netherlands.
Yebusi Garden Place, Tower, 4–20–3 Ebisu, Shibuya-ku, Tokyo, Japan.
Internet: http://www.adobeshop.com and http://www.photodisc.com

Advanced Biotechnologies: Molecular biology kits and reagents, chromosome prints for rapid chromosome identification.

Units B1–B2 Longmead Business Centre, Blenheim Road, Epsom, Surrey, KT19 9QQ, UK.
Wedenstrasse 23, 20097 Hamburg, Germany.
In Japan, distributed by Nikkon Genetics Co. Ltd, Hongso Tsuna Bldg, 17–9, Hongo 6–Chome, Bonkyo-Ku, Tokyo 113, Japan.
Internet: http://www.adbio.co.uk

Agar Scientific: Accessories for microscopy

66A Cambridge Road, Stansted, Essex, CM24 8DA, UK.

Aldrich: see **Sigma**.

Alexis Biochemicals: Distribute a range of products from Affinity BioReagents (polyclonal and monoclonal antibodies), Ancell Europe (human immunology research), Bender MedSystems (immunology), Cayman Chemical, Sigent Laboratories (cell pathology reagents), TaKaRa (cell biology), Neogen Corporation, Toxin Technology, Inc.

3 Moorbridge Court, Moorbridge Road East, Bingham, Nottingham, NG13 8QQ, UK.
In Japan, distributed by CosmoBio Co.
Internet: http://www.alexis-corp.com

Alpha Laboratories: Consumables and plasticware, also supply products from **Vysis, Oncor,** American Diagnostica, and others.

40 Parham Drive, Eastleigh, Hants, SO50 4NU, UK.
Internet: http://www.alphalabs.co.uk

Amersham Pharmacia Biotech: Molecular and cell biology products.

Amersham Place, Little Chalfont, Bucks, HP7 9NA, UK.
800 Centennial Ave, PO Box 1327, Piscataway, NJ 08855–1327, USA.
Otsuka Daiichi Seimei Building, 32–22 Higashi Ikebukuro 2–chome, Toshima-ku, Tokyo 170, Japan.
Internet: http://www.apbiotech.com and, for Cy-series dyes, http://www.amersham.co.uk/life/lcat/nalad/cyedye/index.html

Anachem: Molecular biology products, sells **Gilson** pipettes, **Bio101** distributor.

20 Charles Street, Luton, Beds, LU2 0EB, UK.

Andover Optical: Fluorescence microscopy filters and coatings.

4 Commercial Drive, Salem, NH 03078–2800, USA.
Internet: http://www.andcorp.com/index.html

Applied Imaging: Cytogenetic instrumentation, sell CytoVision and RxFISH color chromosome analysis.

2380 Walsh Avenue, Building B, Santa Clara, CA 95051, USA.
BioScience Centre, Times Square, Scotswood Road, Newcastle upon Tyne, NE1 4EP, UK.
Internet: http://www.aii.co.uk/index1.html

Appligene Oncor: see **Quantum Appligene Lifescreen.**

ASI: Applied Spectral Imaging, sell SkyVision spectral karyotyping systems.

PO Box 101, Industrial Park, Migdal Haemek, 10551 Israel.
2120 Las Palmas Drive, Suite B, Carlsbad, CA 92009 USA.
Neurott Strasse 12, 68535 Edingen-Neckarhausen, Germany.
In the UK, supplied by InterFocus, UK; http://www.interfocus.ltd.uk
Internet: http://www.spectral-imaging.com

Azlon: Plasticware, see **Bibby Sterilin.**

BDH: General laboratory chemicals and supplies (part of **Merck**).

Merck House, Poole, Dorset, BH15 1TD, UK.
Promochem GmbH, Postfach 100955, Mercatorstrasse 51D-46469 Wesel, Germany.
Gallard Schlesinger, 584 Mineola Avenue, Carle Place, New York 11514–1731 USA.
See **Merck** and **Daiichi** for Japan.
Internet: http://www.merck-ltd.co.uk or http://www.chromatography.co.uk

BDI: Imaging software for analysis of fluorescent samples, e.g. MultiFluor imaging software.

Biological Detection Systems Inc. 955 William Pitt Way, Pittburgh, PA 15238, USA.

Becton-Dickinson: Medical diagnostics, labware, DNA extraction alliance with **Qiagen**.

1 Becton Drive, Franklin Lakes, NJ 07417, USA.
Between Towns Road, Oxford, OX4 3LY, UK.
BD Nippon, Akasaka DS Building, 5–26 Akasaka 8–chome, Minato-ku, Tokyo 107, Japan.
Internet: http://www.bd.com

Bibby Sterilin Ltd: Plasticware.

Tillin Drive, Stone, Staffordshire, ST15 0SA, UK.
See **Dynalab** for US distributor.
Internet: http://www.bibby-sterilin.co.uk

BIO 101: Molecular biology reagents.

2251 Rutherford Road, Carlsbad, CA 92008, USA.
In the UK, distributed by **Anachem**.
Funakoshi, Jupiter Uni Building 9–7, Hongo 2–Chome, Bunkyo-ku, Tokyo 103, Japan.
Internet: http://www.bio101.com/index-main.html.

BioCell Research Laboratories: Gold conjugates.

PO Box 78, Ty-Glas Avenue, Llanishen, Cardiff, CF1 1XL, UK.

Biognostik: Antisense technology, clinical diagnostics (Hybridprobes).

Gerhard Gerdes Strasse 19, 37079 Göttingen, Germany.
In the UK sold by **TCS Biologicals**.

Bio-Rad Laboratories: Molecular biology reagents and equipment, confocal microscopes and fluroescence technology.

200 Alfred Nobel Drive, Hercules, CA 94547, USA.
Bio-rad House, Mayland Avenue, Hemel Hemstead, Herts, HP2 4BR, UK.
Bischof Strasse 86, 47809 Krefeld, Germany.
Nippon BioRad Laboratories KK, Japan (several offices).
Internet: http://www.bio-rad.com

Boehringer Mannheim: now **Roche Diagnostics**.

Calbiochem-Novabiochem Biochemicals:

10394 Pacific Center Court, San Diego, CA 92121, USA.
Boulevard Industrial Park, Padge Road, Beeston, Nottingham, NG9 2JR, UK.
MG Tamachi Building 3F, 4–3–7, Shiba, Minato-ku, Tokoyo 108, Japan.
Internet: http://www.calbiochem.com or http://www.cnuk.co.uk

Cambio: sells *STAR*FISH* Chromosome paints (also supplied by **Vysis**).

34 Newnham Road, Cambridge, CB3 9EY, UK.
Internet: http://www.cambio.co.uk

Cambridge Antibody Technology Immunology:

Science Park, Melbourn, Cambridgeshire, SG8 6JJ, UK.
Internet: http://www.catplc.co.uk

Cambridge BioScience: Products and services for life science research, distributes products from Display Systems Biotech, **Molecular Probes**, **Clonetech**, BabCO, Southern Biotechnology Associates, Inc.

24–25 Signet Court, Newmarket Road, Cambridge, CB5 8LA, UK.
Internet: http://www.bioscience.co.uk

CellPath: Cytological preparation products and consumables, Histolene clearing agent.

Cell Path House, PO Box 101, Hemel Hempstead, Herts, HP3 8QE UK.
Internet: http://www.cix.co.uk/~cellpath

Chemische Werke Lowi: Resins, polymers and solvents.

Postfach 1660, D-84469 Waldkraiburg, Germany.
Part of **Great Lakes Chemical** (see under **Great Lakes** for USA address).
Internet: http://www.greatlakeschem.com

Chroma Technology: Design and manufacture optical filters for fluorescence microscopy, and coatings.

72 Cotton Mill Hill, Unit A9, Brattleboro, VT 05301, USA.
Internet: http://www.chroma.com.

Clonetech: Molecular biology.

1020 East Meadow Circle, Palo Alto, CA 94303, USA.
Tulla Strasse 4, 69126 Heidelberg, Germany.
In the UK, distributed by **Cambridge BioScience**.
Internet: http:///www.clonetech.com

Cohu: CCD cameras and digital imaging systems.

PO Box 85623, San Diego, CA 92186, USA.
Internet: http://www.coho.com

CyDye: now manufactured by **Amersham Pharmacia**.

Cytocell: Culture media and probe technology (Chromoprobes).

Somerville Court, Banbury Business Park, Adderbury, Oxfordshire OX17 3SN,
UK.
Internet: http://www.cytocell.co.uk

Daiichi Pure Chemicals: Manufacturers and distributors for, e.g., **New England Biolabs, BDH**.

8–3 Higashinihonbashi, 2–Chome Chuo-ku, Tokyo 103, Japan,
or 13–5 Nihonbashi, 3–Chome, Chuo-ku, Tokyo 103, Japan.

DAKO: Immunology, antibodies and visualization systems.

Produktionsvej 42, DK-2600 Glostrup, Denmark.
Hiraoka Building, Nishinotouin-Higashiiru, Shijo-dori, Shimogyo-ku, Kyoto 600–8463, Japan.
6392 Via Real, Carpinteria, CA 93013, USA.
Denmark House, Angel Drove, Ely, Cambridgeshire CB7 4ET, UK.
Internet: http://www.dakousa.com or http://www.dako.com, and
http://www.dakoltd.co.uk/home.htm

DataDiagnostic: Histoclear (chemical name Limonene, from many chemical suppliers, and also known as Bioclear, Histoclear, Citroclear, etc.). Alternatives such as Neoclear, Merck; Ultraclear, J.T. Baker; or xylene may be used.

Display Systems Biotech: Genomics research tools and services.

1260 Liberty Way, Suite B, Vista, CA 92083, USA.
Lersoe Park Allé 40, DK 2100 Copenhagen, Denmark.
See **Funakoshi** for Japan.
Internet: http://www.displaysytems.com

Dupont: see subsidiary, **New England Nuclear**.

Dynalab Corporation: Plasticware, chemicals and other consumables.

PO Box 112, Rochester, New York 14601–0112 USA.

Enzo Diagnostics: Nonradioactive labeling systems (see also **Roche**, distributor).

60 Executive Boulevard, Farmingdale, NY 11735, USA.
Internet: http://www.enzobio.com

Eppendorf: Microcentrifuge tubes and equipment.

Eppendorf-Netheler-Hinz GmbH, 22331 Hamburg, Germany.
Internet: http://www.eppendorf.com

Fischer Scientific: Instruments and reagents (including **TaKaRa**).

2000 Park Lane Drive, Pittsburgh, PA 15275–1126, USA.

FMC Bioproducts: Chemical labeling kits and instrumentation.

5 Maple Street, Rockland, ME 04841, USA.
Broad Oak Enterprise Village, Broad Oak Road, Sittingbourne, ME9 8AQ, UK.

Fluka: see **Sigma**.

Funakoshi: Distributor, sells products from **Stratagene, Biognostik, Display Systems**.

9–7 Hingo 2–chome, Bunkyo-ku, Tokyo 113–0033, Japan.
Internet: http://www.funakoshi.co.jp.

Genome Systems: Diagnostic services including mapping your clone to chromosomes by FISH. (Subsidiary of Incyte).

4633 World Parkway Circle, St Louis, Missouri 63134–3115, USA.
Internet: http://www.genomesystems.com.

GENSET: Oligonucleotides and probes.

48 Haussler Terrace, Clifton, NJ 07013, USA.
Genset KK, Japan, oligos@po.iijnet.or.jp
Internet: http://www.genset.fr

Genosys or **Sigma-Genosys:** Molecular biology custom services.

1442 Lake Front Circle, Suite 185, The Woodlands, Texas 77380, USA.
London Road, Pampisford, Cambridge, CB2 4EF, UK.
Internet: http://www.genosys.com

GIBCO BRL: Molecular biology products, see **Life Technologies**.

Gilson: Best known for micropipettes (sold by many other companies).

B.P. 45, 95400 Villiers-le-Bel, France.
3000 West Beltline Hwy, Box 6270027, Middleton, WI 53562–0027, USA.
Internet: http://www.gilson.com

Great Lakes Chemical: Resins, polymers and solvents.

500 E 96th St, Suite 500, Indianapolis, Indiana 46240, USA.
See **Chemische Werke Lowi** for European site.
Internet: http://www.greatlakeschem.com

Hamamatsu Photonics: Digital and other camera systems.

325–6,Sunayama-cho, Hamamatsu City, Shizuoka Pref, 430–8587, Japan.
360 Foothill Road, Bridgewater, NJ 08807–0910, USA.
Lough Point, 2 Gladbeck Way, Windmill Hill, Enfield, Middlesex, EN2 7JA, UK.
Internet: http://usa.hamamatsu.com and http://www.hpk.co.jp

Hamilton: Equipment and syringe needles.

PO Box 26, CH-7402 Bonaduz, Switzerland.
PO Box 10030, Reno, Nevada, NV 89520–0012, USA.
Lyne Riggs Estate, Lancaster Road, Carnforth, LA5 9ES, UK.
Internet: http://www.hamiltoncomp.com

Hybaid: Bioanalysis, PCR equipment, hybridization ovens and *in situ* processing machines (subsidiary of Thermo Electron.)

8 East Forge Parkway, Franklin, MA 02038, USA.
Action Court, Ashford Road, Ashford, Middlesex, TW15 1XB, UK.
Internet: http://www.hybaid.com

Invitrogen: Cell biology supplies.

Invitrogen Corporation, 1600 Faraday Avenue, Carlsbad, CA 92008, USA.
Invitrogen BV, De Schelp 12, 9351 NV Leek, the Netherlands.
In Japan, distributed by Funakoshi.
Internet: http://www.invitrogen.com

Jackson ImmunoResearch: Antibodies.

872 West Baltimore Pike, PO Box 9, West Grove, Pennsylvania 19390, USA.
In the UK, distributed by Stratech Scientific Ltd, 61–63 Dudley Street, Luton, Beds, LU2 0NP, UK.
Internet: http://www.jacksonimmuno.com also http://stratech.co.uk

Kaker.com: Microanalysis, microscopy and computer consultancy, web site directory.

Ob Suhi 23, 2390 Ravne, Slovenia.
Internet: http://www.kaker.com

Kreatech: Labeling systems (Ulysis) and other reagents.

PO Box 12756, 1100AT Amsterdam, the Netherlands.
Internet: http://www.kreatech.com

Labworld-online: Equipment on the Internet.

Im Gaisgraben 4, 79219 Staufen, Germany.
Internet: http://www.labworld-online.com

Leica: Microscopes.

Postfach 1120, Heidelberger Straße 17–19, D-69226 Nußloch, Germany.
Leica (UK) Ltd and Leica Microsystems (UK) Ltd., Davy Avenue, Knowhill, Milton Keynes, MK5 8LB, UK.
Internet: http://www.leica.com/mic-sci/index.asp or
http://www.leica-microsystems.com

Leitz: now **Leica.**

Life Technologies: GIBCO-BRL products for cell and molecular biology.

3 Fountain Drive, Inchinnan Business Park, Paisley, PA4 9RF, UK.
Technologie Park Karlsruhe, Emmy-Noerther-Strasse 10, 76131 Karlsruhe, Germany.
8400 Helgerman Court, Gaithersburg, MD 20877, USA.
Internet: http://www.lifetech.com

London Resin Company: Resins for light and electron microscopy embedding (LR White).

PO Box 34, Basingstoke, Hampshire, RG21 2NW, UK.

Merck: Chemicals, see also the subsidiary **BDH.**

Frankfurter Strasse 250, 64293 Darmstadt, Germany.
Hunter Boulevard, Magna Park, Lutterworth, Leics, LE17 4XN, UK.
ARCO Tower, 5F, 8–1, Shimomeguro 1–Chome, Meguro-Ku, Tokyo 153, Japan (see also **Daiichi** for Japan).
Internet: http://www.merck.de or http://www.merck-ltd.co.uk

Molecular Probes: Fluorescent technology.

PO Box 22010, Eugene, Oregon 97402–0414, USA.
PoortGebouw, Rijnsburger weg 10, Leiden, the Netherlands.
In the UK, distributed by **Cambridge BioScience**.
Internet: http://www.probes.com. The book on www.probes.com/handbook/tnotes/tn19.html is particularly useful.

MWG Biotech: Genomic service.

Anzinger Strasse 7, 85560 Ebersberg, Germany.
4170 Mendehall, Oaks Parkway, Suite 160, High Point, NC 27265, USA.
Waterside House, Peartree Bridge, Milton Keynes, MK6 3BY, UK.
Internet: http://www.mwg-biotech.com

National Diagnostic: Histology materials, including Histosol.

305 Patton Drive, Atlanta, Georgia 30336, USA.
Unit 4, Fleet Business Park, Itlings Lane, Hessle, Hull, HU13 9LX, UK.

Novagen: Molecular and cell biology.

PO Box 88641, Milwaukee, WI 53288–0641, USA.
In the UK, distributed by **Cambridge Bioscience**.
In Japan, distributed by **TaKaRa** .
Internet: http://www.novagen.com

New England Biolabs (NEB): Molecular biology reagents and kits.

Knowl Piece, Wilbury Way, Hitchin, Herts, SG4 0TY, UK.
Postfach 2750, 65820 Schwalbach/Taunus, Germany.
See **Daiichi Pure Chemicals** for Japan.
Internet: http://www.neb.com or http://www.uk.neb.com

New England Nuclear (NEN): Cell and molecular biology reagents.

NEN Life Science Products Inc, Boston, MA 02118–2512, USA.
NEN Life Science Products, 1930 Zaventem, Belgium.
DuPont (UK) Biotechnology Systems Division, Wedgewood Way, Stevenage, Hertfordshire, SG1 4QN, UK.
Internet: http://www.nenlifesci.com

Nikon: Microscopes.

6–16, Ohi3–chome, Shinagawa-ku, Tokyo, Japan.
PO Box 222, 1170 AE Badhoevedrop, The Netherlands.
1300 Walt Whitman Road, Melville, NY 11747–3046, USA.
Nikon UK Ltd, Nikon House, 380 Richmond Road, Kingston, Surrey, KT2 5PR, UK.
Internet: http://www.nikon.co.jp or http://www.nikonusa.com

Novocastra Laboratories: Biomedical research products for histopathology, medical microbiology/virology and neuroscience.

Balliol Business Park West, Benton Lane, Newcastle upon Tyne, NE12 8EW, UK.
Distributed in the UK and USA by **Vector Laboratories**.
Internet: http://www.novocastra.co.uk

Olympus: Microscopes.

2 Corporate Center Drive, Melville, NY 11747–3157, USA.
Olympus Optical Co (UK) Ltd, 2–8 Honduras Street, London, EC1B 1HJ, UK.
Internet: http://www.olympus.co.jp or www.olympusamerica.com/microscopes or http://www.olympus-europa.com

Omega Optical: Filter sets for fluorescence applications.

PO Box 573, Brattleboro, VT 05302–0573, USA.
UK supplier: Glen Spectra Ltd, 2–4 Wigton Gardens, Stanmore, Middlesex, HA7 1BG, UK.
Internet: http://www.omegafilters.com

Oncor: see **Quantum Appligene Lifescreen** in Europe (company is split in the USA; see probes sold by **Ventana**): Fluorescent *in situ* hybridization technology and probes.

Internet: http://www.oncor.com gives purchasers of parts of Oncor.

Partec GmbH: Flow cytometers and fluorescent microscope bulbs.

Otto-Hahn Strasse 32, D-48161 Münster, Germany.
Internet: http://www.partec.de

Perkin Elmer Corporation including **PE Applied Biosystems:** DNA sequencing, PCR machines and reagents. (Merged with and see also **PerSeptive Biosystems**).

PE Biosystems, 850 Lincoln Center Drive, Foster City, CA 94404, USA.
Applied Biosystems GmbH, European Life Science Center, Paul-Ehrlich-Strasse 17, 63225 Langen, Germany.
Internet: http://www.perkin-elmer.com

Perceptive Scientific Instruments (PSI): Chromosome analysis systems.

2525 South Shore Boulevard, League City, Texas 75573, USA
Halladale, Lakeside, Chester Business Park, Wrexham Road, Chester, CH4 9QT, UK.
Internet: http://www.persci.com

PerSeptive Biosystems: Fluorescent *in situ* hybridization probes, PCR, DNA sequencing.

500 Old Connecticut Path, Framingham, MA 01701, USA.
Nihon PerSeptive, Zentoku Roppongi Building, 1–7–27 Roppongi, Minato-ku, Tokyo 106, Japan.
Internet: http://www.pbio.com

Pharmacia: see **Amersham**.

Photometrics: see **Roper Scientific**.

PreAnalytiX: Nucleic acid purification systems. See parent companies, **Qiagen** and **Becton Dickinson**.

Promega: Molecular biology reagents.

Promega Corporation, 2800 Woods Hollow Road, Madison, WI 53711–5399, USA.
Promega Ltd., Delta House, Chilworth Research Centre, Southampton, UK.
Promega KK, Tokyo, Japan.
Internet: http://www.promega.com or http://www.euro.promega.com/uk

Qiagen: DNA isolation and cloning kits (see also **Becton-Dickinson** who have an alliance).

Boundary Court, Gatwick Road, Crawley, West Sussex RH10 2AX, UK.
28159 Avenue Stanford, Santa Clarita, CA 91355, USA.
Internet: http://www.qiagen.com

Quantum Appligene Lifescreen: Labeled human painting and oncogene probes and kits.

Salamander Quay West, Park Lane, Harefield, Middlesex UB9 6NZ, UK.

R&D Systems: Immunology and *in situ* hybridization.

614 McKinley Place NE, Minneapolis, MN 55413, USA.
4–10 The Quadrant, Barton Lane, Abingdon, Oxon, OX14 3YS, UK.

Reichert: Microtomes and microscopes. See **Leica**.

Roche Molecular Biochemicals (acquired **Boehringer Mannheim Biochemicals** in 1998).

Roche Dignostics, Sandhofer Straße 116, D-68305 Mannheim, Germany.
Roche Diagnostics UK (Diagnostics & Biochemicals) Limited, Bell Lane, Lewes, East Sussex, BN7 1LG, UK.
Roche Diagnostics Corporation, Biochemical Products, 9115 Hague Road, PO Box 50414, Indianapolis, IN 46250–0414, USA.
Roche K.K., 10–11 Toranomon, 3–chome, Minato-ku, Tokyo 105, Japan.
Internet: http://www.roche.com

Roper Scientific: Scientific grade CCD cameras (formerly Princeton Instruments and **Photometrics**).

PO Box 1192, 43 High Street, Marlow, Bucks SL7 1GB, UK.
Internet: http://www.prinst.com

Scanalytics: Imaging software including IP lab.

Internet: http://www.iplab.com

Scotlab: General supplies and Biovation chromosome probes.

Kirkshaws Road, Coatbridge, Strathclyde, ML5 8AD, UK.

Seshin Pharmaceutical: Digestion enzymes.

9–500–1, Nagareyama, Nagareyama-shi, Chiba-ken, Japan.

Shandon: Cytogenetic laboratory equipment and supplies; Cytospin centrifuges. (Part of Thermo Electron).

171 Industry Drive, Pittsburgh, PA 15275–1034, USA.
Chadwick Road, Astmoor, Runcorn, Cheshire, WA7 1PR, UK.
Japan Tanner Corp., Suita Yakult Building 3F 564, 3–2 Hiroshibacho, Japan.
Internet: http://www.shandon.com

Sigma: Biochemicals, reagents and some equipment. The Sigma family includes Sigma, Fulka, Aldrich, Riedel-de Haën and Supelco.

PO Box 14508, St. Louis, Missouri 63178–9916, USA.
Fancy Road, Poole, Dorset BH12 4QH, UK.
Sigma-Aldrich Chemie, Germany.
Sigma-Aldrich Japan K.K., Tokyo, Japan.
Internet: http://www.sigma.sial.com

Southern Biotechnology Associates: Immunology.

PO Box 26221, Birmingham, Alabama 35260, USA.
In the UK, distributed by **Cambridge Bioscience**.
Internet: http://southernbiotech.com

SPI: see **Structure Probe Inc**.

Stratagene: Innovative technology of life science advancement.

11011 North Torrey Pines Road, La Jolla, CA 92037, USA.
Cambridge Innovation Centre, 140 Cambridge Science Park, Milton Road, Cambridge, CB4 4GF, UK.
Postfach 105466, 6900 Heidelberg, Germany.
In Japan, distributed by **Funakoshi Ltd**.
Internet: http://www.stratagene.com

Structure Probe Inc. (SPI): Microscope laboratory supplies and equipment.

PO Box 656 (569 East Gay St), West Chester, PA 19381–0656, USA.
569 East Gay St, West Chester, PA 19380, USA.
Internet: http://www.2spi.com

Synoptics: CCD cameras and imaging systems.

Internet: http://www.synoptics.com

Taab Laboratories: Microscopy reagents and small equipment.

40 Grovelands Road, Reading, Berkshire, UK.

TaKaRa: Molecular and cell biological products.

TaKaRa Shuzo, 2257 Sunaike, Noji-cho, Kusatsu-shi, Shiga 525–0055, Japan.
Europarc De Barbanniers, 6 Place du Village, 92230 Gennevilliers, France.
In the UK, distributed by **Alexis Biochemicals**.
In the USA, distributed by **Fisher Scientific**.
Internet: http://www.takara.co.jp

TCS Biologicals: Clinical cell biological products, sell Biognostik HybriProbes and Antisense Technology.

Botolph Claydon, Buckingham, MK18 2LR, UK.
Internet: http://www.tcsgroup.co.uk

Techne: Equipment including *in situ* hybridization machines.

743 Alexander Road, Princeton, NJ 08540–6328, USA.
Duxford, Cambridge, CB2 4PZ, UK.
Internet: http://www.techneuk.co.uk

Thermo Bioanalysis (part of Thermo Electron): Holds many important laboratory companies including **Hybaid, Shandon**, Dynex.

504 Airport Road, Santa Fe, NM 87504–2108, USA.
Internet: http://www.thermo.com

Toyobo: Molecular and cell biology reagents, distributor.

17–9 Nihonbashi, Koami-Cho, Chou-Ku, Tokyo 103, Japan.
Internet: http://www.toyobo.co.jp/e/index2.htm

Ushio: Micoscope bulbs.

Asahi-Tokai Bldg, 6-1 Otemachi-2-Chome, Chiyoda-ku, Tokyo 100, Japan.
10550 Camden Drive, Cypress, CA 90630, USA.
Sky Park, Breguetaan 16–18, 1438 BC Oudemeer, The Netherlands.
Internet: http://www.ushio.co.jp or http://www.ushio.com

Vector Laboratories: Immunology and antibody technology, also distribute **Novocastra Laboratories**.

30 Ingold, Burlingame, CA 94010, USA.
3 Accent Park, Bakewell Road, Orton Southgate, Peterborough, PE2 6XS, UK.
Internet: http://www.vectorlabs.com

Ventana Medical Systems: Prelabeled probes, former **Oncor** products.

3865 North Business Center Drive, Tuscos, AZ 85705, USA.
Internet: http://www.ventanamed.com

Vysis: Molecular cytogenetics and genomic disease management products.

3100 Woodcreek Drive, Downers Grove, IL 60515–5400, USA.
Rosedale House, Rosedale Road, Richmond, TW9 2SW, UK, distributed by **Alpha Laboratories**.
Vor dem Lauch 25, 70567 Stuttgart-Fasanenhog, Germany.
Fujisawa Pharmaceutical Co, Ltd, Medical Supplies & Systems Division 4–7, Doshomachi, 3-chrome, Chuo-ku, Osaka 541, Japan.
Internet: http://www.vysis.com

Yakult Pharmaceutical: Digestion enzymes.

1–19 Higashi Shinbashi Minato-Ku, Tokyo 105, Japan.

Zeiss: Microscopes, image analysis, confocal microscopes.

Carl Zeiss Jena GmbH, Zeiss Gruppe, Unternehmensbereich Mikrosopie, 07740 Jena, Germany.
Carl Zeiss Inc., One Zeiss Drive, Thornwood, NY 10594, USA.
Carl Zeiss Ltd, PO Box 78, Woodfield Road, Welwyn Garden City, Herts, AL7 1LU, UK.
Internet: http://www.zeiss.com or http://www.zeiss.co.uk or http://www.zeiss.co.jp

Zymed Laboratories: Methods and material for immunochemistry.

485 Carlton Court, South San Francisco, CA 94080, USA.
In the UK, distributed by **Cambridge Bioscience**.

Abbreviations and definitions

To avoid the need for cross-referencing, many abbreviations are defined in the text. The most commonly used are defined here. A few abbreviations used with *in situ* hybridization, but not in the text, are also defined.

AAF	2-acetylaminofluorene
AP	Alkaline phosphatase
APES	3-aminopropyltriethoxy-silane
BAC	Bacterial artificial chromosomes. Cloning vectors for DNA, typically up to 100 kb long
BCIP	5-bromo-4-chloro-3-indoyl-phosphate
BrdU	Bromodeoxyuridine
BSA	Bovine serum albumin. A protein widely used to out-compete non-specific binding of immunoproteins to protein binding sites.
CGH	Comparative genomic hybridization
CISS	Chromosomal *in situ* suppression hybridization. *In situ* hybridization for chromosome painting with DNA probes and blocking DNA
DAB	3'-diaminobenzidine tetrahydrochloride
DABCO	1,4-diazabicyclo[2,2,2]octane
DAPI	4',6-diamidino-2-phenylindole
DEPC	Diethyl pyrocarbonate
DNase	Deoxyribonuclease. Enzyme degrading DNA
DOP-PCR	Degenerate oligonucleotide primed-PCR
DTT	Dithiothreitol
EDTA	disodium ethylene diamine tetraacetate
EM	Electron microscope (microscopy). Also used as a quality standard for reagents – 'EM grade'
FISH	Fluorescent *in situ* hybridization. ('Fluorescent' is preferred over 'fluorescence' *in situ* hybridization)
GISH	Genomic *in situ* hybridization. *In situ* hybridization using total genomic DNA as a probe
HRPO	Horseradish peroxidase
IRS	Interspersed repetitive sequence
LINE	Long interspersed nuclear element. A class of retroelements
LM	Light microscope (microscopy)
M-FISH	Multicolor FISH
McFISH	Multicolor FISH
NBT	4-Nitroblue tetrazolium chloride
NIB	Nuclear isolation buffer
NOR	Nucleolar organizing (or organizer) region. The major sites of the 45S rDNA loci, at the secondary constrictions on chromosomes
NPG	*n*-propyl gallate
NTE-buffer	NaCl-Tris-EDTA buffer (also called STE-buffer; see Appendix 1)
PBS	Phosphate buffered saline (see Appendix 1)
PCR	Polymerase chain reaction
PI	Propidium iodide

PMSF	Phenylmethylsulfonyl fluoride
POD	Horseradish peroxidase
PPD	2,5-diphenyl-1,3,4-oxadiazol
PRINS	Primed *in situ* hybridization method
PTA	Phosphotungstic acid
PVP	Polyvinylpyrrolidone
rDNA	Ribosomal DNA. Consists of the 5S rRNA genes and intergenic spacers, and a second class, the 45S rRNA genes of the 18S, 5.8S and 25S rRNA genes and intergenic spacers
RNase	Ribonuclease. Enzyme degrading RNA
rRNA	Ribosomal RNA (see rDNA)
SC	Synaptonemal complex
SDS	Sodium dodecyl sulfate, sometimes called lauryl
SKY	Spectral karyotyping
SINE	Short interspersed nuclear element. A relative of retroelement sequences
SSC	Saline (or salt) sodium citrate, or standard saline citrate. A weakly buffered solution of sodium ions, used as an isotonic medium and for control of Na^+ concentration (see Appendix 1)
STE-buffer	Sodium-Tris-EDTA buffer (also called NTE-buffer; see Appendix 1)
TBS	Tris-buffered saline (see Appendix 1)
TdT	Terminal deoxynucleotidyl transferase
TE	Tris-EDTA (see Appendix 1)
TEM	Transmission electron microscope (microscopy)
v/v	Volume added to volume, so 60% v/v acetic acid is 60 ml glacial acetic acid and 40 ml water
w/v	Weight added to volume, so 10% SDS is 10 g SDS powder added to water to make a final volume of 100 ml
YAC	Yeast artificial chromosomes. Cloning vectors grown in yeast for DNA typically 100–500 kb long

Nucleotide base abbreviations:

R	A, G – puRine
Y	C, T – pYrimidine
M	A, C
K	G, T
S	G, C
W	A, T
H	A, T, C – not G (letter after G: h)
B	G, T, C – not A (b)
V	G, A, C – not T (u, v)
D	G, A, T – not C (d)
N	A, T, C, G – aNy

Nucleotides: dCTP, dATP, dGTP, dUTP, TTP

Index